Health & Safety Inspections

How to avoid them, how to deal with them

Calgarth House • 39-41 Bank Street
Ashford • Kent
Tel.: 01233 653500 • Fax: 01233 647100
customer.services@indicator.co.uk • www.indicator.co.uk

We have made every effort to ensure that the information contained in this book reflects the law as at 1st September 2002.

The male pronoun has been used throughout. This is simply to avoid cumbersome language and no discrimination or bias is intended.

ISBN 0-9543737-1-5

Introduction

As you're only too aware, running a business in today's climate means juggling conflicting demands. Dealing with the authorities that have responsibility for enforcing health and safety may be a new experience for you. In fact, one of the most common problems is a misunderstanding of an inspector's powers.

The aim of this book is to provide a useful tool to help you avoid an inspection and to deal with the authorities if a visit becomes inevitable.

The book is divided into two main sections. The first section looks at how the local authorities and the Health & Safety Executive (HSE) operate. Although the two agencies have enforcement powers over different types of premises, the rules governing their operation are the same. However, if there are any differences between the two, such as in the complaints procedure, this is mentioned. The second section covers the Fire Brigade, which has jurisdiction over all workplaces, irrespective of the type of business being operated.

This book can be used both as a source of reference that you can dip into as and when required and also as a workbook to help you plan and deal effectively with visits. Most importantly, tips are given to help you so you don't end up spending money unnecessarily.

Helen Rideout

Editor

Indicator Advisors & Publishers

Table of contents

Section I. Health and Safety Enforcement Authorities

1 Overview

2 Who does the inspections?

3 Why will they visit?

4 Inspection techniques

5 What are they looking for?

6 Inspectors' powers

7 The inspector's wish list

Section II. Fire Authorities

1 Overview

2 Who undertakes inspections?

3 Selection criteria

4 Powers of entry

5 The inspector's wish list

6 Problems encountered (and how to resolve them)

7 Enforcement action and appeal

Section I

Health and Safety Enforcement Authorities

This section provides guidance to help you prepare for and deal with any health and safety inspections that may occur. It's based on the experiences of those who have enforced health and safety in a wide range of premises, as well as those who've been on the receiving end of such inspections.

1. Overview

You won't find continual references to specific pieces of legislation in this section. Instead, references have deliberately been kept to a minimum, as the purpose is to describe what you must do within the context of the powers that inspectors have when they visit. The approach given is very much "no frills" and only states what would be expected from a small business in terms of compliance, hence the emphasis on a practical approach.

This section also provides information on how premises are selected for inspections and visits, through to the powers of inspectors and your rights if things go wrong. It also describes in some detail, the system used by local authorities, to assess how often they will visit. This is crucial in helping you plan for the future.

Much prominence is given to paperwork and its importance in health and safety management. This isn't just so you can keep your inspector happy by giving him the equivalent of a small rain forest to plough through. It's actually so that you can show you've done what the law requires. Even basic paperwork might well help to establish a sound "defence".

2. Who does the inspections?

2.1. WHO DOES WHAT?

2.1.1. Local authority inspectors

Local authority inspectors cover the following workplaces: retailing from small shops to department stores, some warehouses, most offices, hotels and catering establishments, sports, leisure, consumer services businesses and places of worship. So in general, their activities are concerned with the "softer" workplaces.

2.1.2. HSE inspectors

HSE inspectors have responsibility for organisations that are any of the following: factories, building sites, mines, farms, fairgrounds, quarries, railways, chemical plants, offshore and nuclear installations, schools, hospitals and other places where there is work activity. In general, HSE inspectors tend to be associated with the higher risk premises.

2.2. WHO WILL VISIT YOU?

There's some overlap in enforcement between the two, but as a rule of thumb, your enforcement authority will depend on the main activity that's undertaken on your premises. For example, if your business is as a wholesaler with a service bay for repair and maintenance of vehicles, you'll come under the remit of the local authority as your main activity is wholesaling. Conversely, if your main activity is a manufacturing factory but you have several offices on site, the HSE will be your enforcement agency.

2.2.1. Different types of inspectors

If you come under the jurisdiction of the local authority, there may be some confusion as to the type of inspector that may turn up on your doorstep. Whilst most people have heard of Environmental Health Officers (EHO's), many businesses may be unaware of the existence of the Technical Officer (TO).

Whilst the majority of inspections are likely to be carried out by an EHO, a TO may visit instead or accompany the EHO. They tend to have different qualifications, though it's equally important to stay on their good side and not make comments about wanting to be inspected by a "proper EHO". This has been known to happen and doesn't make for a happy or relaxed visit. But, irrespective of whether your inspector is an EHO or a TO, their powers are given to them under the same legislation as for HSE inspectors.

This legislation requires that inspectors must have "suitable qualifications" to enforce all legislation relating to safety. This can mean that a TO is only authorised to undertake some enforcement duties and not the full range, although in practice there's rarely any difference. When you're being visited, you can check their powers by asking to see their authorisation warrant. This sets out the enforcement duties that they are authorised to carry out and tells you if they have the authority to come onto your premises uninvited. In practice, no inspector will be sent out who doesn't possess the authority to conduct an inspection.

2.3. THEIR APPROACH

The point of the visit is to see that health and safety standards are acceptable and that risks are being properly managed on your premises. Although you may be tempted to think otherwise, it's not part of their remit to try and catch you out, or to make your life difficult.

2.4. INSPECTORS' BEHAVIOUR

Although the following is arguably a gross generalisation of behaviour, your inspector may fall into one of the following categories in respect of his approach to enforcement and advice giving.

2.4.1. The professional approach

This is your ideal inspector as his approach is one that aims to manage risk effectively. He won't quote legislation at you, but instead will dip into it in order to explain things or demonstrate exactly what's required.

If you're really lucky, this type of inspector is somewhat pragmatic in that they have an appreciation of the commercial realities that you have to operate under and will keep them in mind when offering you any advice.

Don't hesitate in asking inspectors what they would do in your situation. They have a vast range of knowledge, so use it.

2.4.2. The bureaucratic approach

Those with bureaucratic tendencies are the opposite of the professional and are happiest when buffered within the comfort zone of laws and rules. In fact, he (and it's usually a "he") is happiest when quoting rules and subsections to you. As a rule, these inspectors tend to be older, having spent much of their working lives in enforcement, and still aren't really comfortable with the risk-based approach. If you get a visit from one of these inspectors, don't hesitate in drawing the conversation back to the realities that you face with your staffing or budget levels. Make the most of their knowledge and turn the conversation round to ask them what they would do.

2.4.3. The commercially unaware approach

This type of behaviour in an inspector is fairly rare, which is fortunate. They tend to have similar tendencies to the bureaucrat in terms of emphasising compliance with legislation, but they also lack commercial awareness. This is likely to be because there's no requirement for inspectors to have had prior experience of working commercially within health and safety. Whilst this often doesn't present a problem, it has the potential to do so in terms of the advice that they may offer you, which you may feel is unrealistic or is too costly for you to justify.

3. Why will they visit?

3.1. VISIT, WHAT VISIT?

3.1.1. The phone call

You may have been going about your business quite happily for the last year or more, only to receive a letter or telephone call from your local inspector announcing that they "would like" to visit your premises for an inspection. The words "would like" are somewhat deceptive as, like death and taxes, a visit at some point is quite likely and isn't something that you have the option to decline. The best you can hope for is to try and postpone it for a few weeks, because you're just too busy at the moment.

However, never rely on this as a tactic as neither authority has to agree to a postponement. This is partly because the inspectors, especially those from the local authorities, are themselves subject to performance standards and must complete inspections within a given time frame. The authorities also have the power to enter your premises unannounced, at all reasonable times. So if you're really unlucky, you may find that an inspector just turns up during business hours without any warning.

In general, it's impossible to give hard and fast rules about inspectors' visiting habits, as there's as much variety within a single area as there is between the HSE and a local authority. Some will ring up and make an appointment with you, whilst others will just appear on your doorstep and the approach will vary from inspector to inspector.

3.1.2. First impressions

If an inspector turns up unannounced and is greeted by an untidy reception area, you have little chance of getting rid of him and re-arranging the inspection.

Never forget that if an inspector suddenly turns up on your doorstep, your success at persuading him to come back at a later date will rely on him being happy with what he sees in your entrance. First appearances are crucial and if an inspector looks around and sees a clean and tidy environment, you'll have a much better chance of postponing the visit. But if he's greeted by an untidy environment with fire extinguishers propping doors open, don't expect to get rid of him in a hurry.

Another tactic that may be employed is to ask you for five minutes of your time in order to give him a feel for the way you manage your premises.

Remember that your inspector will have heard every trick in the book as to why a visit should be postponed. He may agree to yours, but don't rely on it.

From your point of view this is a much better deal than having a full inspection foisted on you, so you agree forgetting that you have little choice in the matter. But be warned that in five minutes an inspector can cover a lot of ground and if he doesn't like what he sees, he'll do a full inspection on the spot.

3.1.3. Helping yourself

Suppose you have an inspector in your entrance or reception area who has talked his way in to having five minutes of your time. As fate would have it, he's caught you at a bad time and you know that your premises aren't at their best. If you can try and guide him to a nearby area of your business that is tidier for a quick discussion then do so. Alternatively, if your entrance area is one of the tidiest parts of your business, your best bet is to try and keep him there. He also likes paperwork and this is irrespective of which enforcement agency he belongs to. So if you can distract him for a few minutes by showing him your accident books and drawing his attention to your health and safety poster and employer's liability insurance certificate on the wall, you may succeed in getting away with just five minutes, on this occasion at least.

3.2. CHOOSING WHOM TO VISIT

You may think that it's divine intervention behind an inspector's choice to visit your premises, but in reality a number of very different factors combine to determine when your number is up. But don't forget, most premises are likely to be inspected at some point, so don't approach it from the viewpoint that you're being singled out. In reality, inspectors utilise a range of information sources when deciding whom they're going to visit and these are as follows:

3.2.1. Database

One major source of information is the records that inspectors already hold on businesses and organisations in their area. These will be in the form of paper files or more commonly, computer files on establishments that they've already visited or had dealings with. If you've reported an accident to either the local authority or HSE in the past, your details will certainly be on their database.

3.2.2. Local knowledge

Like anyone else, inspectors are aware of their surroundings and take a keen interest in new office or industrial developments that are built in their patch. Similarly, they notice when premises change hands, so this is an informal way for them to be aware of the organisations in their area.

3.2.3. Area surveys

Local authorities carry out area surveys whereby each borough is divided up into wards. Each ward may be surveyed street by street to see what businesses are there. It's not known how frequently these surveys are updated, as it's likely to vary between authorities and will depend to some degree on their resources.

3.2.4. Registration of business

This is for those businesses that come under the remit of the local authority. Registration with the council is required shortly before you open a new business or at the time you start to employ people. The exception to this is if you are a small family business only employing close family. If your business is an office, retail shop, wholesaler or a caterer (e.g. pub, takeaway or restaurant), then you'll need to contact your local authority to obtain and complete **Form OSR1** to register. These forms only require basic information but contain enough detail for inspectors to be able to find their way to your door.

If your business is a factory, it's a legal requirement that you're supposed to give at least one month's notice to the HSE before the occupation or use begins.

3.2.5. Food registration

If your business involves the storage, sale, production or preparation of food, it must also be registered with your local authority. This will mean that you're likely to receive visits from the food hygiene team as well as more general visits for health and safety purposes.

3.2.6. Complaints by workers or public

This is probably the worst way possible to come to the attention of your local authority. Sometimes a disgruntled employee or visitor to your premises may make a complaint direct to the authority about some health or safety defect. If inspectors learn of your existence this way, you can be assured of a visit.

3.3. ANY OTHER FACTORS

To some extent the likelihood of a visit from your local inspector depends on what the current hot topics in enforcement are. Luckily these are in the public domain, but it helps if you know exactly where to look. The biggest clue can be found in the "**HSC Strategic Plan 2001-2004**", which is a free publication from HSE Books (telephone number 01787 881165).

It details a number of key areas of both health and safety and specific industries that both the HSE and local authorities will be concentrating on over a three-year period. Under the current strategic plan, it divides the eight chosen hot topics (priority programmes) up and emphasises different aspects of each during the three-year period. These eight priorities are shown in the table below.

Health issue	Safety issue	Industry
Musculoskeletal disorders	Falls from heights	Construction
	Slips and trips	Agriculture
Stress	Workplace transport	Health services

With the exception of health services these are discussed below.

Call 01787 881165 for a free copy of the HSC's "Strategic Plan 2001-2004". It tells you what inspectors are concentrating on – meaning you can focus your response accordingly.

3.3.1. Health concerns

Occupational health is receiving a much higher profile than previously and the Strategic Plan reflects this. The areas include musculoskeletal disorders such as lower back problems, shoulder and neck pains and wrist problems. These can arise in all industry sectors from offices through to assembly line work. Another key area and one that's the subject of 2002's European Week on health and safety is stress. Enforcement interest will cover all areas of work-related stress including bullying and violence at work. Respiratory problems continue to occupy the agenda and asthma and asbestos remain key issues. The following examples are taken from the current Strategic Plan and illustrate how the inspectors will target different workplaces.

<u>Musculoskeletal disorders</u>

In a typical year 9.9 million working days are lost due to work-related musculoskeletal disorders such as back pain. As a result, the HSE was scheduled to have visited 3,290 workplaces in 2001/2002 on inspections to investigate how employers manage manual handling activities. Local authority inspectors will focus their resources on handling in the food and drink industry and on patient-handling in care homes.

Work-related stress

Stress was chosen because it now accounts for about 20% of all reported cases of occupational ill-health every year. This is leading to 6.5 million working days being lost to employers annually. However, instead of the direct enforcement approach that is being used for musculoskeletal disorders, the approach to stress will be different. The focus is on the development of guidance and occupational health programmes for employers. What may happen though is that inspectors may ask you about your approach to stress management when they visit you for other reasons. It's always possible that a more direct enforcement approach may appear in the next Strategic Plan from 2005 onwards.

3.3.2. Safety concerns

Safety priorities include slips, trips and falls which affect all workplaces to some degree and workplace transport that will affect far fewer. Other topics more associated with heavier industries are health problems arising from excessive noise and hand/arm vibration (HAV) from over-use of tools like drills.

Slips, trips and falls

This has become a priority as figures show that in the local authority enforced sector, slips and trips account for about 42% of major injuries to employees and half of all non-fatal injuries to members of the public in 1998/1999. As a result local authority inspectors are conducting a series of investigations and targeted inspections on higher incidence premises. Whilst in 2002/2003, it's the turn of HSE inspectors to make similar inspections in their areas.

Falls from heights

This was chosen because falls from heights cause 80 fatalities a year and around 5,600 major injuries per year. The HSE plan to make 960 inspections during this period, whilst local authorities will target off-site workers such as window cleaners. Targeted visits and blitzes will continue to be made in the 2002/2004 periods.

Workplace transport

This was picked because 70 people die every year in workplace transport-related accidents and there are around 6,000 injuries every year that lead to people being off work for three days or more. As a result both enforcement agencies have run a series of workplace blitzes that will continue to run into 2004. These blitzes are likely to affect organisations with loading bays and warehouses.

3.3.3. Industries

Construction

Not surprisingly, enforcement priorities are concentrated on those industries that present the highest risk in terms of particular safety hazards. Special emphasis is being given to the construction industry with inspectors making a series of inspections to buildings sites all over Great Britain. These blitzes have been well publicised in the national press and another series is about to start. One blitz in the north of England and Scotland saw enforcement action being taken against nearly half of all sites visited.

Agriculture

Agriculture has one of the highest incidence rates of fatalities in any industry sector. A series of targeted inspections is being carried out which looks at workplace transport, falls from heights and musculoskeletal disorders. These inspections will carry on into 2003, along with conferences and safety awareness days.

3.4. TYPE OF VISIT

Once they know of your existence, an inspector's visit to your business may be prompted by a number of factors, over which you'll have either some or absolutely no control.

3.4.1. Accident or incident investigations

This will be an unplanned visit by an inspector in response to the submission of an accident report form or a phone call about an incident that has occurred on your premises. The incident could be down to the reporting of an occupational disease, but that is quite rare as many employers, including large ones, are unaware of their duties in this area. But it would apply where an employee has been diagnosed with a complaint such as asthma or dermatitis attributed to their working environment. At the end of the visit, the inspector will give a verbal explanation as to what has to be done.

3.4.2. Planned inspection

This is the most common type of visit and as mentioned above it may be by appointment or via an inspector turning up on your doorstep. The purpose is to review your ability to effectively manage health and safety so that it isn't a risk to staff, contractors, visitors or members of the public.

3.4.3. Complaint

Sometimes a person will contact the authorities and make a complaint about your premises. It may be an employee, a visitor or anybody who has witnessed something of concern to them e.g. someone visiting a restaurant may complain about food poisoning or unhygienic practices, or a member of the public may have concerns about the way scaffolding has been erected on a building site.

3.4.4. Follow-up visit

This will take place after an inspector has made a visit in order to deal with either a complaint or to investigate a serious accident. At present, this visit may be more likely to occur if your business comes under the jurisdiction of the local authority rather than the HSE. Although in future, both agencies will be making more visits. The purpose of this visit will be to see whether you have implemented the actions and/or recommendations made by the inspector on the last visit.

3.4.5. Sweep/blitz

This can take the form of either an area blitz or an industry blitz. In the case of an area blitz, manpower resources will be allocated so that a rolling programme is put into place that will see every business within a certain area visited by the authorities. If it's an industry blitz that's being undertaken, usually high-risk industries such as construction or agriculture are targeted. Sometimes a combination of the two takes place.

Industry blitzes are a favourite of the HSE and are usually accompanied by good press coverage e.g. the HSE has undertaken a massive construction blitz throughout the country that has seen many construction companies receiving Prohibition Notices. These prevent them from carrying out certain activities until health and safety improvements have been made.

4. Inspection techniques

4.1. THE DIFFERENT ELEMENTS

Both local authority and HSE inspectors employ a range of inspection techniques. Much depends on the inspector's own style and the reasons behind the visit. In all cases though, the visit will consist of a combination of at least three of the four techniques described below, with the formal interview reserved for when a business is potentially in serious trouble.

Although it would be officially denied, it's not unknown for inspectors to walk around a business's car park in order to see what cars the directors drive. This isn't just confined to medium or large undertakings and its purpose is to give an inspector an idea of the true wealth of a company. So if you want to plead poverty as to why you can't implement an inspector's recommendations, having a brand new BMW or Jaguar in the car park, won't help in convincing him. If you know the inspector is due, drive your spouse's sensible hatchback to work that day!

4.1.1. Observation

Inspectors will notice far more than you'd ever realise from simple observation. As with anything that's familiar, you're used to your workplace and there are probably many things that you don't notice or that you take for granted, e.g. a piece of carpet that's torn or a typist's chair that's broken.

When an inspector visits he will be looking at a number of things. The most obvious will be the state of the housekeeping. This will be the case whether it's an EHO inspecting an office or an HSE inspector walking around a factory. Other obvious areas will be how clear the walkways are and whether there's safe access. From an inspector's point of view, something is safe or it isn't. So if a walkway is blocked with boxes, junk or trip hazards, it isn't safe and won't be until all offending items are removed. If the route is a fire escape, then it should be kept completely clear, but in other areas, maintain a width of one metre along corridors that's clear of boxes, but don't block up doorways.

With safe access, an inspector will look for the less obvious such as how areas are accessed for maintenance work. This will include areas like the roof where access is often required for cooling towers and plant rooms.

Look at maintenance schedules and walk around your own premises. This'll flag up any problems, so you can say that you've identified the risks.

It'll also include areas that are used to access areas for lifts and ventilation. Unfortunately, it's easy to forget about having good safe access for these areas as they're not regularly used and are often out of bounds to the majority of staff.

4.1.2. Informal interview

This will often take place whilst walking around the premises and the nature of the questions will depend on the premises being visited. So an HSE inspector walking around a factory will ask questions about how a particular item of machinery works. He may then start asking more detailed questions about which safety features the machine has. Related questions could include enquiries as to training records, test certificates and how any lifting issues are dealt with. Likewise an EHO visiting offices may ask about the lighting and office space if workers look cramped.

4.1.3. Formal interview

If there's a serious breach of health and safety on your premises, an inspector may wish to hold a formal interview. The purpose of this will be to see if there's enough evidence to justify bringing enforcement action. This type of interview may also be used following a serious accident on your premises. More information is given in Chapter 6.

4.1.4. Document review

As there are so many statutory requirements for different forms of paperwork, expect to be asked to present a variety of documents to your inspectors. Again, the exact nature will depend on the workplace but the following are a safe bet in any business. These are; the safety policy, risk assessments, portable appliance testing records and assessments undertaken for those who work with display screen equipment, such as computers. In other premises expect to be asked for tests on local exhaust ventilation (LEV), statutory lift tests and paperwork for your cooling towers.

5. What are they looking for?

5.1. AN INSPECTOR'S POINTS SYSTEM

Whether you've been on the receiving end of an inspection before, or are about to have your first, you're probably wondering how you will be assessed. Luckily, the new era of more open government makes it easier to find out, as long as you know where to look and how to interpret the information. For some strange reason, the authorities aren't exactly proactive in drawing your attention to this information...

5.1.1. Local authority approach

Local authority inspectors use what's known as an "Inspector Rating System" (IRS) to fix the frequency of planned visits. In other words, it's this system that will determine why you only have one visit every few years, whilst your neighbouring business may receive two visits a year. Extracts from this IRS are given below and throughout this chapter. The full version can be found on the following website: www.hse.gov.uk/lau/lacs/67-1.htm.

5.1.2. HSE approach

It's only recently that HSE inspectors introduced their own inspector rating system, which is similar to that used by the local authorities. It's known as the **Enforcement Management Model** and whilst the principles of it are widely applied by inspectors, it's not often formally followed. Neither is it meant to be a procedure in its own right. Because of this, this chapter concentrates on the IRS, which local authority inspectors do use for inspections. If you follow this, you should keep your HSE inspector happy, as after all, their system is based on the HSC Enforcement Policy, which will apply equally in HSE enforced sectors.

5.1.3. How the system works

The IRS comprises seven elements, each of which should be rated on a scale of 1-6, with 6 representing the worst situation. The elements are **safety hazard, health hazard, safety risk, health risk, welfare, public risk and confidence in management**. With the public risk element, it's possible to score zero, if you don't have the public on your premises.

The system then applies weighting factors (multipliers) to each rating element to reflect their relative significance. So the highest weighting is applied to public risk and confidence in management, whilst the lowest is applied to welfare. The current weighting factors are:

Rating element	Weighting
Safety hazard	6
Health hazard	6
Safety risk	9
Health risk	9
Welfare	5
Public risk	10
Confidence in management	10

The HSE has a free pro-forma form for risk assessments that can be obtained via www.hsebooks.co.uk ***or by ringing the HSE on 01787 881165.***

5.1.4. Small businesses

The rating system allows inspectors to exercise some discretion with small businesses as most or all management functions may lie with one individual. Unfortunately, there's no definition of what counts as a small business, so an inspector will look at each business individually. This discretion means that inspectors can allow for the fact that in many cases procedures may not be as well documented as they would like. Where this applies, they will be looking to see how you operate in practice.

5.1.5. What are they looking for?

Confidence in management

When assessing confidence in management, inspectors will ask to see your paperwork. This is discussed in far more detail in Chapter 7. However, it will include a review of your safety policy, risk assessments, consultation with your employees over health and safety issues and training. An inspector will also look at your previous enforcement history and your awareness of the health and safety risks from your business.

Your inspector will then make an assessment on these elements to give you a score. If he feels that your safety policy is so limited as to be useless, or that your safety management is non-existent, you will already be on your way to a maximum score of 6 under this section.

Safety hazard

The definition of a safety hazard is the potential of an activity, a method of work, or a piece of machinery etc. to cause harm. This is wide ranging and looks at the level of safety hazard to everyone in the workplace except members of the public, who are assessed separately.

The nature of any safety hazards will depend to some degree on the nature of your business e.g. factories and food premises will usually have safety hazards on a level not found in the majority of offices. The presence of particular hazards such as gases and machinery won't in themselves influence the rating, as the key is how you manage them.

The following table shows how safety hazards are scored in relation to the injuries they can cause. The scores are on a scale of 1-6; with negligible risks scoring 1 and major risks scoring 6:

Output	**Examples of harm occurring**
Negligible	Minor injuries needing first aid.
Low	Other injuries unlikely to lead to permanent disability, but may need time off or hospital treatment.
Low - Medium	Injuries unlikely to cause permanent disability but are likely to need more than three days off work and/or over 24 hours in hospital.
Medium - High	Fractures (not fingers or toes), amputations. Long-term or permanent disability likely.
High	Extremely serious injury/fatality.
Major	Multiple fatalities.

Many small businesses could expect to be in the negligible to low category, except those that are factory based and have the potential for more serious injuries.

Health hazard

This is defined as the potential of a substance, noise or a method of work etc. to cause harm. The rating value given should be representative of the single most serious hazard present and no account is taken of the standard to which the health hazard is controlled. Equally, no account is taken of the number of people who may be exposed to the different health hazards at a workplace. Inspectors only include hazards under the control of the occupying business and not those posed by other businesses, but which may affect those working for you.

Health hazards cover a wide range from those that can be found in offices such as poor lighting, ventilation and repetitive strain injury (RSI) to those more connected with the facilities management side of a business. This area is broad and includes everything from the rooftop downwards. Therefore the existence of asbestos will be a major consideration to a number of businesses. It's often said that health hazards are less tangible than safety hazards, but inspectors give them equal importance, so you need to be prepared.

The following table shows how health hazards are scored on a scale of 1-6, with negligible risks scoring 1 and major risks scoring 6:

Output	Examples of hazards
Negligible	Non-hazardous substances: Noise levels less than 80 decibels (dbA).
Low	Noise levels less than 85 dbA. Use of irritant substances, like bleach. Some manual handling.
Low - Medium	Noise levels over 100 dbA for a short time or 90 dbA usually. Use of harmful or corrosive substances e.g. some types of oven cleaner. Potential for upper limb disorders.
Medium - High	Use of toxic substances, use of skin or respiratory sensitisers e.g. some water treatment chemicals and some hairdressing products.
High	Very toxic substances e.g. CO_2 used in confined spaces, asbestos or legionella.
Major	This category is reserved for premises not enforced by local authorities.

Safety risk

This is defined as the likelihood that the harm from a single particular hazard will be realised. Whilst the single most serious hazard should be the primary focus of the inspector, the risk rating may be increased in some cases to reflect the cumulative effect of a number of significant hazards e.g. premises with both flammable and explosive chemicals on-site. The assessment should be limited to considering the adequacy of control of immediate risks i.e. those found at the time of the inspection.

The hazard rating is an important factor in deciding on the risk rating and it's possible for a high safety hazard to warrant a low risk rating if it's well managed. Equally it's possible for poor standards in managing safety risks to lead to a higher scoring. Inspectors will also assess the levels of exposure to a particular hazard, based on the numbers exposed and the duration of the exposure. Safety risks are scored on a scale of 1-6, with negligible risks scoring 1 and major risks scoring 6.

Health risk

This is defined as the likelihood that the harm from the single most serious hazard will be realised. Whilst this should be the primary focus of the inspector, the risk rating may be increased in some cases to reflect the cumulative effect of a number of significant hazards e.g. exposure to asbestos whilst working without adequate protection in a joinery workshop. The assessment should be limited to considering the adequacy of control of immediate risks i.e. those found at the time of the inspection.

The hazard rating is an important factor in deciding on the risk rating and it's possible for a high health hazard to warrant a low risk rating if it's well managed. Equally it's possible for poor standards in managing health risks to lead to a higher scoring. Inspectors will also assess the levels of exposure to a particular hazard, based on the numbers exposed and the duration of the exposure e.g. a large hairdressing salon where staff often handle perming solutions and toners etc. is likely to present higher health risks than a small barber's shop where the owner occasionally uses perming lotion. Health risks are scored on a scale of 1-6, with negligible risks scoring 1 and major risks scoring 6.

Welfare

This one is self-explanatory and refers to the standard of welfare provision in a particular workplace. The facilities provided should be commensurate to the needs of those in the workplace and the rating should reflect the inspector's opinion of the adequacy and standard of facilities provided. For example, if your workers are engaged in manual work that causes them to get dirty, you'll be expected to provide showers and changing rooms for their outdoor clothes, whereas if your environment was office based, showers would be a nicety, but not actually necessary.

The table overleaf shows how welfare is scored and is somewhat subjective:

Rating score	Output	Examples
1	Very good	Excellent welfare provision, no deficiencies noted.
1	Good	Good standard of welfare provision, only minor deficiencies noted.
3	Fair - good	Reasonable standard of welfare provision, some deficiencies noted, but none serious.
4	Fair - bad	Reasonable standard of welfare provision, but some serious deficiencies noted.
5	Bad	Poor standard of welfare provision.
6	Very bad	Absence or very poor standard of welfare provision.

Public risk

This is defined as the likelihood that the public will be harmed by activities in a workplace. This includes those not employed in connection with your business and customers and passers-by. As with health and safety risk, public risk depends on the hazards present, the extent that they are controlled, the numbers of people exposed and the frequency of the exposure. Inspectors should also take into account vulnerable groups such as children and the elderly who may be at greater risk due to lack of appreciation of risk or to infirmity. In reaching a rating score, inspectors must ignore common hazards such as electric shocks and slips, trips and falls, as they are widespread, both in and outside a workplace.

High-risk premises would be regarded as those with unsafe systems of work and deliberate non-compliance.

5.1.6. Adding it all up

Once a score has been derived from adding up each of the scores that make up the seven elements of the rating system (confidence in management, welfare, etc.), there are some other factors that are then used, namely something called “elapsed years” and a weighting based on national accident data for different types of premises within the local authorities’ remit. This is where the scoring system clearly departs from the system used by the HSE, as part of it only relates to extra scores levied on local authority enforced premises.

5.1.7. Elapsed years

Premises/work units are divided up into three broad groups with "A" representing the greatest risk and "C" the least risk. An extra score of +15 is added annually to premises in groups B1-4, but not to premises placed into group C. Premises that are placed into group A are deemed to be high-risk and they will be visited annually, so the elapsed years factor doesn't come into effect.

5.1.8. National accident data

The scoring system also includes an element that takes into account national accident levels in the main sectors that come under local authority enforcement. This information is updated annually and is based on the accident reports that local authorities receive. This extra score is weighted to reflect the higher rate of accidents that occur in catering and retail premises over those found in offices.

5.1.9. Hazard/risk categories

When the elapsed years factor and national accident data scoring has been added on, a final score can be calculated. This score will then determine which of the hazard/risk categories your premises will be placed in and this will determine the frequency of your visits. So if you're in group A, you can expect annual visits, whilst those in group C could be visited once every five to ten years. As mentioned above, there are three broad categories: A, B and C, with B being sub-divided into four sub-groups as follows:

Description	LAC Group	Score
Highest hazard/risk	A	Greater than 186
Intermediate hazard/risk	B1	171 - 185
	B2	156 - 170
	B3	141 - 155
	B4	126 - 140
Lowest hazard/risk	C	Less than 125

5.2. DRAWING CONCLUSIONS

Enforcement priorities should always be given to the highest risk premises that reach a minimum score of 186 or above. So it follows that the lowest hazard/risk premises won't be a priority in terms of visits.

However, inspectors will keep group C premises under constant review to ensure that they should still remain within this group.

5.2.1. Monitoring low-risk premises

They do this by use of their local knowledge and from other information such as planning applications made to other council departments. After all, if a planning application shows that there's going to be a period of construction work, then immediately a low-risk, group C premises will become higher risk due to all the new hazards and contractors being introduced into it. Other factors will include accident reports and if the council start to receive details about reportable accidents, then a review and inspection are likely to be automatic. Inspectors may stay in contact with group C premises by use of seminars, telephone calls or mailshots.

5.2.2. Environmental health issues

In case you were wondering about the impact of other potential environmental health issues such as noise and food hygiene on the visit, don't worry. Other environmental health functions like these are deliberately excluded when rating premises for health and safety risks. Therefore, a full health and safety at work inspection should normally take place at intervals solely determined by the IRS. However, if an inspector makes a visit due to a public health issue or an outbreak of food poisoning from the work's canteen and he notices a significant health or safety issue, he should deal with it.

5.3. YOUR RIGHTS

5.3.1. Ask away

You have the right to ask an inspector for particular details of the inspection or to ask about anything else. This acts as a good check on any local authority inspectors who may have the tendency to be over zealous or if there's personal friction between you.

5.3.2. Right to complain

If the inspector doesn't cooperate, your next option is to approach his manager for details of the inspection. In the unlikely event that the manager refuses to play ball, each local authority has a complaints procedure. Another option is to approach a councillor and then the Local Government Ombudsman for your region. Although it's very rare, judicial review and private legal action are also possible. In practice, it would be considered unreasonable not to inform you why you were being visited at certain frequencies.

There are three Local Government Ombudsmen in England:

Tony Redmond

Local Government Ombudsman

21 Queen Anne's Gate

London SW1H 9BU

Tel: 020 7915 3210

Fax: 020 7233 0396

Covers: London boroughs north of the River Thames (including Richmond, but not Harrow), Buckinghamshire, Berkshire, Hertfordshire, Essex, Kent, Surrey, Suffolk, East Sussex, West Sussex and Coventry City.

Patricia Thomas

Local Government Ombudsman

Beverley House

17 Shipton Road

York YO30 5FZ

Tel: 01904 663200

Fax: 01904 663269

Covers: Birmingham City, Cheshire, Derbyshire, Nottinghamshire, Lincolnshire and the north of England (except York and Lancaster).

Jerry White

Local Government Ombudsman

The Oaks No. 2

Westwood Way

Westwood Business Park

Coventry CV4 8JB

Tel: 024 7669 5999

Fax: 024 7669 5902

Covers: London boroughs south of the River Thames (except Richmond and Harrow), York, Lancaster and the rest of England.

There are separate Local Government Ombudsmen for Scotland (Tel: 0131 225 5300) and Wales (Tel: 01656 661325).

6. Inspectors' powers

6.1. POWER OF ENTRY

6.1.1. Day or night

An inspector's authorisation gives him the power to enter premises at all "reasonable" times. This generally means that he can visit and legally demand entry during business hours, so at the very least, this will be between 0900 and 1700 weekdays. However, he does have the power to enter at any other time if he believes that a dangerous situation exists. This is very unlikely but may happen if an employee or member of the public makes contact to inform him of a problem that the emergency services are either unaware of, or may be in the process of attending.

6.1.2. What about property damage?

Although it's a very unlikely scenario, if a forced entry is needed, there could potentially be issues as to payment for any damage to the premises due to the break-in. However, whether or not the outcome goes in your favour, will depend on the circumstances of each case. In particular, on the reasonableness of the break-in and on the outcome of any case brought against the inspecting authority. It would be safe to assume that where any inspector has reasonable cause, and he normally does, no compensation can be claimed.

6.2. POLICE ASSISTANCE

An inspector has a right to be accompanied by a police officer if he believes that he will be obstructed or prevented from entering premises. This would certainly be a black mark against you and is guaranteed to land you in big trouble before the inspector even sets foot in the door. Inspectors may request that the police attend with them in the unlikely event of them needing to access premises after hours, but they aren't duty bound to do this.

6.3. BRINGING A COLLEAGUE

An inspector is also at liberty to be accompanied by anyone else authorised by the local authority such as a specialist in a particular field e.g. an electrical specialist to review electrical fittings or wiring. He may also bring any equipment or material onto the premises that he needs for the purpose of his visit.

So if you find yourself in this unfortunate position, grit your teeth and be as cooperative as possible. Although it will depend very much on the circumstances giving rise to the visit, discuss any concerns or problems that you may have with him, e.g. that the visit may cause undue disruption to staff productivity.

6.4. INSPECTING THE PREMISES

Once he has entered, the inspector has a legal right to be able to inspect the premises and carry out the investigation undisturbed. This may mean that nobody else has the right of access to a particular area whilst the investigation is being carried out.

6.4.1. Business interruption issues

Naturally these decisions could have serious business interruption issues. If an inspector believes that there's a risk of serious personal injury then he may serve a Prohibition Notice, which will take effect immediately. You could attempt to negotiate with him over the detrimental effects that such a Notice may have on the business, but the decision will lie with the inspector and his managers. Not surprisingly, there's limited case law on the subject, but the examples given below suggest that for public policy reasons, the need to protect health and safety will always override concerns about business interruption.

> <u>*Example 1*</u>
>
> ***Harris v Evans and Another*** *(1998)*
>
> *This case shows that an inspector isn't liable in negligence for economic damage caused to a business through the issuing of Notices. In other words, you can't claim against an inspector or his enforcing authority for any financial loss you may suffer due to his mistake.*
>
> *Harris (H) ran a mobile bungee jumping business and was advised by an HSE inspector that he should comply with an Approved Code of Practice issued by the Standard Association of British Bungee. He started business in 1992 on the strength of that advice.*

In 1993, he was visited by an HSE inspector who advised him that the mobile crane shouldn't be used until it was certified as being fit for bungee jumping. He followed this advice and two local authorities served Improvement and Prohibition Notices on him for his trouble. Due to intervention by the Secretary of State for Employment, these Notices were later withdrawn (an unusual move) on the basis that the advice given by the HSE inspector wasn't in line with current HSE policy.

H claimed compensation from both the inspector and the HSE for the economic loss to his business caused by the Improvement and Prohibition Notices. However, the claim failed for several reasons. The most relevant one was that in giving the advice that led to the issuing of the Notices, the HSE didn't owe a duty of care to the owner of the business affected.

The second point is that H didn't appeal against the Improvement and Prohibition Notices, which in hindsight was a mistake as matters could have been resolved far sooner, with less financial loss to his business.

Example 2

Grovehurst Energy Ltd v Strawson *(1990)*

This case shows that prospective loss of profit and an undertaking to apply extra safety precautions isn't sufficient grounds for the suspension of a Prohibition Notice.

Grovehurst (G) generated and supplied electrical power and steam to three paper manufacturers. If it failed to supply this power, the loss to the manufacturers was £86,000 per day. Unfortunately, a Prohibition Notice was served on G as the equipment used to generate the power was deemed unsuitable for further use. G appealed against the Notice saying that extra safety precautions were being taken to prevent injury in the event of equipment failure. The appeal was rejected because there was a very great risk of injury should the equipment fail.

6.5. A FULL INVESTIGATION

The investigation may involve taking measurements, photographs, and samples, or any other recording that an inspector may need for the purpose of an investigation. Likewise, he has the authority to require that something be dismantled for the purpose of an examination or test, e.g. a piece of machinery. His powers also include the right to request facilities such as office space, a telephone and to ask for any help he may need.

6.6. INTERVIEWS

As part of the investigation, inspectors may interview and ask questions of anyone they consider appropriate and you have no right to stop them. In fact, any attempt to do so will reflect very badly on you and may be taken as an attempt to hamper the investigation.

6.6.1. The interview process

If an interview takes place on-site, it will usually be done in a room provided by you. A standard form tends to be used for the interview and statements may be taken. However, where a person has been questioned, his or her answers aren't admissible as evidence in any future proceedings. At the end of the interview, the interviewee must sign a declaration as to the truth of their answers.

6.6.2. Interviews under PACE

PACE stands for the Police and Criminal Evidence Act 1984. If an inspector wishes to conduct an interview under this Act, it will be because the enforcing authority is seriously considering prosecuting you. Due to the gravity of these interviews, which are recorded, an individual who's authorised to speak on behalf of the company must be present, e.g. a director. A list of likely questions is usually sent in advance, along with the invitation to the interview.

A solicitor should be present for any interview under PACE as it removes any opportunity for the interviewee to later turn around and say that they weren't properly represented, thus potentially invalidating the evidence.

6.6.3. Perils of non-cooperation

Sometimes, employers try to prevent their staff from cooperating with inspectors and this seems to be most prevalent in small firms. A typical scenario is where the owner of a business works on-site and his employees either feel threatened or intimidated by them or have a sense of loyalty to their employer and don't want to say the wrong thing. Unfortunately, on a scale of one to ten, this one will score you top marks for shooting yourself in the foot!

6.6.4. Hand-holding colleagues

As well as being able to question whomever they choose, inspectors can exclude others from the interview, which is a very effective way of dealing with difficult employers. However, the person being questioned does have a right to have someone of their choosing present with them. Although it depends very much on the nature of the interview, the choice of individual may be a lawyer.

6.7. CRIMINAL SANCTIONS

It's a criminal offence for anyone to try and prevent an inspector from carrying out his powers. So this means that hampering any investigation in any way e.g. by trying to prevent inspectors talking with employees or by failing to answer their questions honestly may lead to prosecution.

7. The inspector's wish list

7.1. PAPERWORK

You can almost guarantee that as soon as you have your inspector sitting down in your office with a coffee, he will ask to see the following paperwork:

7.1.1. Health and safety law poster

These posters cost less than £10 and can be ordered from any health and safety supplier.

They are a legal requirement and the recent advertising campaign is because the format of the poster has changed. What's more, the law requires that they must be prominently displayed on your premises and if you have more than one site, one is needed for each. An alternative to the poster is to distribute leaflets to each of your employees that cover the same information as contained in the poster. The majority of employers find using the poster is much easier.

Since 1 June 2002, you have to include on the poster details of those who have particular responsibilities for health and safety within your business. It may only seem like a poster, but the law says that you can be prosecuted for not having one of these on display. It's more likely that an inspector will just give you a verbal ticking off and expect you to order one.

7.1.2. Employer's liability insurance

Keep a copy of your employer's liability insurance certificate on a wall where it is prominent. Ensure that it's current.

You should be familiar with this as it's compulsory for covering your backside in the increasingly litigious society that we live in. Its purpose is to cover you for any liabilities to your employees that may arise out of injury or disease from their employment with you. However, what you may not be aware of is that the law actually requires that there should be a copy of your employer's liability certificate in a prominent position on each site of your workplace.

7.1.3. Accident book

The purpose of the accident book is to record all accidents. As a minimum, accident books must include the following information: full name, address and occupation of injured person, date, time and place

Inspectors sometimes ask staff and contractors where the books are located, so ensure that they know where they are.

of accident, cause and nature of injury and details of the person making the entry in the accident book.

Inspectors often ask to see these books, so they should be kept in a prominent place, such as a reception area or site office. Staff should be told where they can find them. Sometimes inspectors ask employees where they are located, so make sure that your employees and any temporary staff and contractors know where to find them.

7.1.4. Procedure for reporting accidents

Inspectors will expect to see a proper procedure in place for reporting accidents, as having books on the premises isn't enough. A system doesn't have to be complex, as the main aim is to be able to show compliance with the reporting requirements. Therefore, it's often easier to make the manager of a particular area or activity responsible for ensuring that accident reports are made and that you're aware of staff sickness arising from any accident. This is important as the nature of the accident may mean that an employee is away from work for three days or more and it becomes reportable. If this happens, or if a visitor of yours is taken to hospital that must also be reported to your local authority. But don't worry; inspectors won't be prosecuting you for making the odd report over the ten-day period, as long as you don't make a habit of it.

7.1.5. Safety policy

If you employ five or more people an inspector will ask to see your health and safety policy. This document is a management tool that states your general policy on health and safety at work and the organisation and arrangements in place for putting it into practice. There will be a number of things that an inspector will check to see are included. These are as follows:

<u>The health and safety policy statement</u>

This goes at the very beginning of the policy and is normally a few paragraphs setting out your commitment to health and safety.

<u>Responsibilities</u>

It's important that the policy document sets out the responsibilities for health and safety within your business. Again this will vary, but it should include a section on the person or group of people, such as a board of directors, who have overall responsibility for health and safety. There should also be some detail on what these responsibilities include.

Health and safety risks

The policy should document any known health and safety risks that could arise from work activities. It should also include what they are and any action that's needed to remove or control them. So if the work processes include using chemicals, then the control measures that have been put in place should be given e.g. looking to substitute with less hazardous chemicals. Finally, this part of the policy should clearly state who is responsible for controlling the risks and how often they will be reviewed. It should also include who has responsibility for carrying out risk assessments.

Consultation with employees

This is a legal requirement and an inspector will be looking to see how this is managed. If the workplace is unionised, there'll need to be a list of the union safety representatives and in a non-unionised workplace, this will need to be the appointed employee representatives. The policy should also state how regular the meetings are and how workplace consultation takes place in practice. Don't be surprised if an inspector asks you for minutes of any meetings.

Safe plant and equipment

This section should state who's responsible for identifying when maintenance is needed, who draws up maintenance procedures, who to report problems to and who purchases new equipment.

Safe handling and use of substances

If hazardous substances are used on the premises, the policy should state those who are responsible for undertaking and reviewing any COSHH assessments done under the **Control of Substances Hazardous to Health Regulations 1999**.

Information, instruction and supervision

This should provide information such as where the Health and Safety poster is displayed and who supervises and trains new recruits and young workers. It should also include details of who gives induction training, job specific training and who in the organisation keeps training records.

Accidents and first aid provision

This will give basic details on where any first aid equipment is stored and the names of any first aiders or appointed persons on site. It should also give the name of the person who reports accidents to the authorities (where applicable). If the nature of the work requires that health

The HSE has a new free leaflet "Stating your business: Guidance on preparing a health and safety policy document for small firms." This can be downloaded at www.hse.gov.uk/pubns/indg324.pdf.

surveillance must be given e.g. where employees work with substances hazardous to health such as certain chemicals, then the names of those responsible for arranging and keeping records of any health surveillance must also be given.

Monitoring

This section names those responsible for monitoring working conditions and safe working practices and will often be the manager responsible for a particular area. It should also name those responsible for investigating workplace accidents and any work related sickness.

Emergency procedures

This should state who has responsibility for putting together emergency procedures such as for fire safety, and what the evacuation procedures are. It should also mention the names of those who are responsible for checking escape routes, fire extinguishers and alarms.

7.1.6. Staff training records

Keeping records of training and what it includes can help cover your back in the event of a claim being made against you.

The records should include the name of the person having had the training, a summary or checklist of what the training includes and a rough date of refresher courses if required. Keeping this information is also useful in the event of an accident or a claim as defending your position is always easier if you have proof that an employee has received training on a particular process or piece of equipment.

7.1.7. Risk assessments

Unless your premises consist only of offices and are modern, an inspector is almost guaranteed to ask to see your risk assessments. These have been a legal requirement for nearly ten years, so expect comments if you haven't conducted any. However, if you've not had an inspection before, or they happen infrequently, an inspector is most likely to give you a time period in which to get the main ones done.

If you know that an inspection is approaching, concentrate on getting the most important assessments done and use the HSE pro forma to help you.

7.1.8. Display screen assessments

If you have office workers, you may be asked for copies of Display Screen Equipment (DSE) assessments. This is another type of risk assessment, but is specifically concerned with the immediate working environment of those who use computers regularly in their jobs. You're

If your workstations aren't compliant, having DSE assessments will make your inspection that bit easier and buy you some time. But, be prepared for a follow-up visit to ensure that you've made the necessary changes.

less likely to be asked for these assessments if an inspector is happy with what he sees when looking at how your office staff work.

7.1.9. COSHH assessments

COSHH assessments are a legal requirement under the **Control of Substances Hazardous to Health Regulations 1999** and an inspector will expect to see assessments where such chemicals are used in the workplace. Your biggest clue as to whether an assessment is needed is to look at the label on the bottle to see what markings it has. If it contains any given in the following table then an assessment will be needed.

Chemical type	Description
Very toxic	Skull and crossbones
Highly flammable	Flame
Dangerous to the environment	Dead tree and fish
Explosive	Explosion
Oxidising	Clear circle with flames on top
Corrosive	Chemical pouring on to hand
Harmful/irritant	Black cross

A new HSE website** www.coshh-essentials.org.uk **is free and will do the assessments for you. All you need to do is input information on what is used in your workplace and for what purpose.

A common mistake is to confuse the material safety data sheets (MSDS) that a manufacturer or supplier of chemicals sends you with a COSHH assessment. The purpose of the MSDS is to describe the known hazards of the chemicals and to give information on how they should be handled, stored and disposed of. They will also provide information on what to do in the event of an emergency, such as a fire and what first aid measures are needed.

7.1.10. Maintenance paperwork

The volume of paperwork will depend very much on the nature of your organisation, but even for office-based businesses, expect to have to produce items such as electrical certificates and written evidence of portable appliance testing. Other paperwork of interest to inspectors includes records of any statutory inspections that have taken place such as for extract ventilation, lifts or machinery.

If you use equipment such as hoists for any lifting activities, then there's a legal requirement for regular maintenance and testing. If you have lifts or boilers on the premises, evidence of inspection by a "competent person" will be expected. This means someone who has sufficient training, experience or knowledge and will often be an external engineer, due to lack of in-house knowledge.

Inspectors can tell much about a business from the state of its sanitary facilities as in badly maintained premises they tend to be neglected.

7.2. GOOD SANITARY AND WELFARE FACILITIES

7.2.1. Toilets

Although it may sound like an unusual question, you may be asked how many toilets you have on site as there are even laws governing their provision. A table to help see if you comply is given below.

An inspector will want to see hot and cold running water, soap and facilities to dry hands. Obviously, any facilities provided should be in a good state of repair and regularly cleaned.

No. of people at work	No. of toilets	No. of washbasins
Mixed Use		
1-5	1	1
6-25	2	2
26-50	3	3
Men only		
1-15	1	1
16-30	2	1
31-45	2	2

7.2.2. Changing rooms

If your work activity requires your employees to wear a uniform or overalls or other clothing as part of their job, then you must provide changing rooms. An inspector will expect to see that they are clean and well maintained with seating and with means to hang clothes e.g. with hooks or pegs. Wherever possible, separate changing facilities for men and women or separate use of the same facilities are recommended.

7.2.3. Drinking water

It's a legal requirement to provide drinking water. It should ideally come from the public water supply as a primary source although bottled water dispensers are perfectly acceptable as a secondary source.

7.2.4. Eating facilities

If facilities are available for eating food, inspectors will want to see certain standards of hygiene and cleanliness e.g. a regular regime for cleaning surfaces and equipment. Although it sounds obvious, the location is important and needs to be somewhere hygienic.

7.2.5. Smoking

There's no specific law against smoking at work, though you are required by existing legislation to protect the health of your employees and others affected by smoking. Many employers have responded to this by either imposing an outright ban on smoking at work, or by introducing designated areas where employees can smoke. It is the effectiveness of this arrangement that an inspector will be interested in. So if you have designated smoking areas and the smoke affects employees in other areas of the building, possibly due to inadequate extract ventilation, an inspector may pick up on it.

7.2.6. First aid provision

Any first aid facilities provided, or areas used for first aid, should be clean and each location should have at least one first aid box that's kept well stocked. There's no legal requirement as to what first aid boxes should include, but as a rule of thumb an inspector would expect to see an assortment of the following: different sized plasters, a couple of sterile eye pads, bandages, safety pins, medium and large sterile wound dressings and a pair of disposable gloves.

First aid boxes shouldn't include certain items such as painkillers or aspirin as certain people are allergic to them and first aiders aren't competent to prescribe drugs.

The requirement for first aiders as opposed to having an "appointed person" will depend on the size and nature of the premises and the number of sites. Inspectors may want to see evidence of how organisations have assessed the risks on their own particular premises taking into account numbers of staff, the risks of the work that's carried out and the distribution of staff within the premises. Therefore an office based organisation occupying one site, with 25 staff will have very different needs to an industrial premises employing a similar number spread over different locations and several minutes walk from each other.

7.3. THE PREMISES

7.3.1. Electrics

Inspectors will expect to see a workplace that doesn't have trailing cables everywhere as they are trip hazards. So if nothing else can be done, like rearranging desks and electrical equipment to more convenient electrical sockets, buy some mats from a safety or office equipment supplier.

In terms of paperwork, you'll be expected to have available records of portable appliance testing (PAT) unless all your electrical equipment is very new.

7.3.2. Lifts

If you have lifts on the premises, expect to be asked for paperwork such as evidence of maintenance and test schedules. This has to be done by a competent person (a specialist lift engineer) every six months if the lift carries people and every year if it's only for goods. So if you know that an inspection is due or that a visit could be likely, ensure that you have a regime in place. If you're stuck, your insurance company should be able to help you find an accredited company capable of undertaking inspections to the required standard.

If you're in premises owned by a landlord, ask to see evidence of maintenance schedules for your own peace of mind as well as being able to show an inspector that you're being pro-active. This always makes a good impression and it's free.

7.3.3. Lighting

Not surprisingly, the law has a preference for natural light over artificial light. So it's better for people to work at ground level or above rather than in a basement office. Where artificial light is needed and its use is inevitable, certainly on dull days and during the winter months, it needs to be sufficient to allow staff to carry out their tasks.

Cover your back by asking your window cleaner to give you a "method statement". This describes how he or she undertakes the job on your premises, following an identification of the hazards e.g. difficult access.

7.3.4. Slips, trips and falls

An inspector will want to see that the flooring in your workplace is in good condition and isn't uneven with holes or bumps. If there are problem areas, then he will want to see that adequate precautions have been taken to protect against accidents e.g. through the use of barriers or safety tape. Special account needs to be taken of those with impaired vision or no sight.

Surfaces of floors and any traffic routes, which are likely to get wet or be subject to spillages, e.g. a factory floor or catering premises, should be of a type that doesn't become too slippery. If a risk assessment deems it to be necessary then a slip resistant coating should be applied. Any floors next to machinery such as for woodworking or grinding should be slip resistant and be kept free of any slippery substances or trip hazards.

7.3.5. Window cleaning

An inspector may show an interest in how the window cleaning is undertaken and will certainly do so if your premises are at or above first floor level and/or access is difficult. Where this is so, safety harnesses may need to be used and these will require anchorage points to be secured into walls.

7.4.UPPER LIMB DISORDERS

7.4.1. Manual handling assessments

The HSE has produced useful guidance "Getting to grips with manual handling for small firms." It can be downloaded from www.hse.gov.uk/pubns/indg143.pdf ***or by telephoning HSE Books on 01787 881165.***

Unless your organisation is solely office based, an inspector may ask you for copies of your manual handling assessments. This applies where catering staff, porters, delivery staff, warehousemen and production line staff are employed, to name but a few. They will show how you have assessed the risks that may arise from any activities requiring lifting, pushing, pulling, carrying or moving an object. This isn't just because the weight itself might be heavy, but because the job activity involves twisting and stooping, which can present its own dangers such as straining back muscles. Also work that is done in confined or restricted space and at height needs to be assessed as the nature of these environments can add extra hazards to manual handling activities.

8. What happens next?

8.1. AT THE END OF THE VISIT

The usual course of action at the end of a visit is to give you advice on how best to comply with the law, or on how best to achieve standards of good practice. Minor areas may be dealt with by the inspector verbally explaining what is required and providing you with guidance where necessary. However, the success of this depends very much on the relationship between you and him. So the better the relationship, the better the outcome will often be.

8.2. FOLLOWING THE VISIT

Within a couple of weeks of the visit, you'll be sent a "Confirmation of Inspection" form that will indicate any areas where there are problems. This letter will identify areas where improvements should occur and will highlight those areas where improvements are recommended. Any points marked "recommended" don't have to be done by law.

If matters are more serious, you're likely to receive a letter and Notice(s) setting out any work which **is** necessary for you to carry out in order to comply with statutory requirements. You may either receive an Improvement Notice or a Prohibition Notice.

8.2.1. Improvement Notices

An inspector will issue an Improvement Notice where standards are lacking, say where there's no hot water provided or where statutory safety training hasn't been carried out. A very common area is where either inadequate or no risk assessments have been undertaken for DSE. The Notice will set a date by which time the work required must be completed.

Failure to comply with the date given is very likely to result in a prosecution for both the original offence and for failure to comply with the Notice. If you find yourself in this situation, ensure that you receive a risk assessment from the inspector. This is because there's case law to say that before serving an Improvement Notice, an inspector must carry out some form of risk assessment.

It would be very rare for an inspector to be able to close down premises completely unless there were a risk of imminent danger e.g. poorly managed demolition work, or a building about to collapse.

Example

***Kwik Save Group Plc** (1993)*

An EHO served an Improvement Notice on Kwik Save, stating that a serious accident had been reported that involved stock knives. The Notice stated that the firm had to replace all stock knives with safety knives. Kwik Save appealed to an employment tribunal.

The tribunal held that the Notice should be quashed as the EHO had made too little use of her powers under S.20 of the Health and Safety at Work etc. Act 1974. In particular she had failed to properly assess the risks arising from the continued use of the stock knives. Therefore, the firm wasn't in breach of its obligations.

8.2.2. Prohibition Notices

These normally take effect from the moment they're issued, where there's a risk of serious personal injury. Therefore, the commercial implications of receiving one of these Notices should never be underestimated. Like Improvement Notices, failure to comply with the Prohibition Notice is likely to result in prosecution, with stiffer penalties, due to the risks of the breach being potentially more serious.

When issuing a Notice, the inspector doesn't actually have to state what remedial measures are necessary in order to comply. So this is likely to be one of those occasions where having a good relationship with him could really be to your benefit.

Example

***MB Gas Ltd v Veitch** (1991)*

Veitch was an EHO who had inspected a petrol station following a complaint about the unsafe storage of liquid petroleum gas (LPG). During his inspection, he discovered breaches of two HSE Guidance Notes and as a result the LPG was moved outside the building in which it had been stored. Unfortunately, the EHO still believed that the LPG posed a risk of serious personal injury, so he issued a Prohibition Notice. The Notice didn't specify the nature of the breach in exact terms and the owner of the petrol station alleged that the Notice was defective because of this.

It was held that the Notice was to stand as it was and the inclusion of details relating to the nature of the breach was an option and not an obligation. Besides, Veitch had told the owner of the petrol station what was required.

8.2.3. Objecting to a Notice

This can be done either before or after a Notice has been issued. Guidance from the Health and Safety Commission requires an inspector to give advance warning of his intention to issue an Improvement Notice. This two-week period gives you the opportunity to discuss the problem and any remedial action that you may be able to take before the Notice is issued.

8.2.4. Prosecution

This is usually as a result of an accident or a total disregard for the law or deliberate non-compliance such as a fatality or serious injury caused by lack of maintenance. Sometimes the hazard has been identified but the control measures removed such as guarding on machinery or where previously identified asbestos has been removed in a dangerous manner. Prosecutions can also occur for failure to comply with a Notice. A fine of up to £20,000 and/or six months imprisonment can be given as the penalty. The decision to prosecute must go through senior managers and must comply with the prosecution policy of that enforcement agency e.g. is it in the public interest to prosecute, or does the seriousness of the offence merit a prosecution?

In Scotland the Procurator Fiscal decides whether a prosecution should be brought and this may be because the enforcing authority has recommended it. However, the Procurator Fiscal has the authority to investigate the circumstances and institute proceedings independently of an enforcing authority.

8.3. COMPLAINING ABOUT AN INSPECTOR

8.3.1. An EHO

In the unlikely event that this is necessary, making a complaint is quite a simple process. If the issue can't be sorted out with the inspector concerned, you can contact his manager and ask for your complaint to be investigated. If you're still not satisfied, you can then use the local authority's formal complaints procedure.

8.3.2. An HSE inspector

You can either speak to or write to the inspector's manager who'll investigate your complaint and keep you informed as to what's happening. The majority of complaints are settled this way and often immediately. If you're still not satisfied you can write to the Director General of the HSE. At the time of writing, the current Director General is Jenny Bacon. You may also write to your MP to ask them to take up your complaint with the HSE.

8.4. APPEALS

If you're unlucky enough to be on the receiving end of either type of Notice, you'll be told in writing of your right of appeal to an employment tribunal. This process is explained on the back of the Notice itself and gives information on how to appeal and within what period it may be brought. The form on which any appeal is to be made is included with the Notice and is called "Notice of Appeal". Any appeal must be lodged within 21 days of the issue of the Notice.

In the case of an Improvement Notice, any remedial action required by it is suspended whilst the appeal is pending. Notices are usually issued after the inspector goes back to the office and they are thoroughly checked by their managers for content and correctness.

9. Accident investigations

9.1. THE ACCIDENT INVESTIGATION PROCESS

If there's an accident connected with work and an employee or a self-employed person is either killed or suffers an injury that keeps them away from work for three days or more, you must report it to the authorities. Likewise, if a member of the public e.g. a visitor or customer to your premises is killed or taken to hospital that must also be reported. You can notify the authorities by phone as long as the correct form is sent to either your own enforcement agency or the Incident Contact Centre (ICC) within ten days of the accident or incident occurring. If you use the ICC, the form can be completed online at www.riddor.gov.uk, faxed on 0845 300 9924 or the information can be given over the telephone on 0845 300 9923.

9.1.1. Will I be investigated?

Depending on the nature of the accident and who your enforcement agency is, you may end up getting a visit from your local inspector to investigate the accident. At present your chances of a visit will largely depend on your particular area as both HSE and local authority enforcement activity have dropped in recent years, with local authority health and safety inspections having fallen by 25% in the last five years. However, this has been picked up and will gradually begin to improve following a new revised enforcement policy from the Health & Safety Commission (HSC).

This means that both enforcement authorities have tended to target their resources at the higher risk activities within their sectors. So local authorities will probably concentrate on light industrial and catering operations, whilst the HSE will concentrate on the industry sectors with the highest rate of injuries e.g. construction, transport, storage and communication, manufacturing and agriculture.

9.1.2. What can you expect?

These visits tend to be pre-arranged, as the inspector will want to see certain information and may wish to interview key individuals, so you should have some time to get yourself prepared. You need to carry out your own investigation before the inspector arrives. He will need to look at a number of areas which will form the basis of any meeting you have. If you incorporate the following into your own investigation, it will go a long way towards having all the information you need available at the time of the inspection.

Witness statements

These should be taken as soon as possible if anyone witnessed all or part of the accident. Don't underestimate the importance of this, especially with insurance claims, as the statement of one witness can make all the difference e.g. if an employee deliberately disabled safety features on machinery and the witness was the only one who saw them do it. Ensure that any witnesses sign and date statements to avoid complications later on.

Get the witness to sign and date any statement made.

Photographs and drawings

With some accidents it may be helpful to take photographs of the accident site e.g. where something has fallen from a height or where part of a structure has collapsed.

Information on equipment

If a piece of equipment has caused or contributed to the accident, an inspector may want to see the specification for it or the instruction manual. So dig them out and have them ready for the visit.

Risk assessments

If the accident was caused whilst someone was following a process at work, you're likely to be asked for any risk assessments that have been done. So if a secretary's chair broke whilst she was typing and she injured her back you'd be asked for a DSE assessment. Alternatively, if the injury happened to someone like a porter who was lifting as part of his job, expect to be asked for a manual handling assessment. If the accident happened to a visitor who slipped in your reception area, an inspector may ask about similar accidents and whether a risk assessment had been done on slips, trips and fall hazards in your workplace.

If you don't have risk assessments in place, depending on the circumstances of the accident, you may "get away" with being given a set time period in which to do them. However, if the accident is very serious and the investigating authority decides to prosecute, it can include a failure to undertake a risk assessment in the case against you.

Training

An inspector may want to see your training records and an outline of training given, depending on the type of accident that occurred. This is almost a certainty for accidents involving machinery, specialist vehicles or even kitchen equipment in a restaurant.

Your recommendations

Even if it means doing the risk assessment after the accident, it's important to have something to show the inspector.

By the time the inspector arrives, he will expect you to come up with some ideas on how to prevent a recurrence of the accident. So if the accident were a slip on some flooring that led to an employee being off work for a week, you'd be expected to have looked at the flooring itself to see if it was fit for its purpose e.g. safety flooring where spillages are likely.

Assuming that the floor was suitable, you'd be expected to look at the way you deal with spillages i.e. is the floor regularly checked, who's responsible for the area and are spillages cleaned as soon as they're identified? Another factor would be the cleaning regime and a look at the cleaning agents used.

With some accidents, you may genuinely have little idea as to exactly why they happened e.g. machinery failure, but as long as you can demonstrate that you've made an effort to get to understand the cause, e.g. by contacting the machine's supplier or the manufacturer for advice, an inspector should be a useful source of help to you.

If an inspector doesn't visit your premises, you may find that you get a phone call to discuss the accident and in particular, what your plans are to prevent a repeat. In this situation, your proposed recommendations are still important, as they may be what stands between you getting just a phone call, or a phone call leading to a visit if the inspector isn't satisfied with your response.

9.1.3. The visit

Taking the above points in mind, the following examples show what an inspector will be looking to see on a visit to investigate a typical accident that's occurred in two very different workplaces.

Example 1

A trainee chef who slipped over on some oil whilst walking out of the kitchen.

An inspector may wish to do the following:

- *A risk assessment for slips, trips and falls in the kitchen.*
- *The specification for the flooring.*
- *The cleaning regime e.g. how often is the floor cleaned and what's used to clean it?*

- *Cleaning schedules that show the frequency of cleaning.*
- *Any policy on slippages e.g. getting kitchen staff to wipe spillages when noticed.*
- *Details of any training the chef has received.*
- *Speak to any witnesses and see their statements where relevant.*
- *Talk to the chef or any other members of kitchen staff.*
- *Take extra statements.*

Example 2

Crushing injury due to cleaning machinery without guarding in place.

An inspector may wish to do the following:

- *To see the working environment and the machine.*
- *Look at the guarding to see why the machine could operate without it.*
- *Speak to any witnesses and see their statements where relevant.*
- *Talk to and take statements from the injured and any witnesses.*
- *Take photographs, measurements and make drawings.*
- *Look at machinery specification.*
- *Ask questions as to how the machine works.*
- *Look at underlying causes e.g. pressures on employees to clean without guarding to reduce downtime.*
- *Speak to supervisors and managers about the accident to gain an idea as to culture.*
- *Look at accident books to see if there've been similar accidents before.*

9.2. INSURANCE CONSIDERATONS

Unfortunately in today's world, the saying "where there's blame, there's a claim" means that there may be a claim sitting on the back of any accident that's reported to the authorities. So one minute you may be breathing a sigh of relief that the authorities aren't following it up and the next, a solicitor's letter lands on your desk. Unfortunately, much of the battle has come down to the quality of your paperwork and the "no-win, no-fee" solicitors will often play the game of asking you for a million policies and procedures. This is a deliberate tactic to deter you and your insurers from defending a claim. So this is another reason why a really

good system for investigating accidents can help you and keep your claims down by giving your insurers enough ammunition to make it worth defending a case.

10. Enforcement – the future

10.1. NEW HSC ENFORCEMENT POLICY

A new policy outlining when and how action will be taken to investigate and prosecute for breaches of health and safety in the future has been issued. This concerns the investigation of workplace accidents and when the decision to prosecute should be taken.

10.1.1. Investigating workplace incidents

In future both enforcing authorities will have to examine several factors when deciding whether or not to investigate a workplace accident. This will include looking at your previous health and safety record and your previous enforcement history, e.g. if you have previously been prosecuted or been given a Notice.

10.1.2. The nature of the accident

This will assess the severity and scale of the potential, as well as actual harm. Another consideration is the seriousness of any potential breach of the legislation. These factors will then need to be assessed in conjunction with the authorities' current enforcement priorities. So if the accident was in the current HSC strategy, it's more likely to be investigated.

10.2. WHEN SHOULD THERE BE A PROSECUTION?

In an attempt to get consistency between and within the enforcing agencies themselves, the policy has set out the circumstances when a prosecution should take place. These include the following:

10.2.1. Workplace death

Where a death occurs due to a breach of health and safety legislation, a prosecution should now follow. Although many employers would assume that this would automatically be the case, it hasn't been in the past.

10.2.2. Recklessness

A prosecution should occur where there has been a reckless disregard of health and safety requirements. This could happen in a number of scenarios such as employees working at heights without fall protection or safety harnesses, or workers stripping out asbestos without the qualifications or personal protective equipment to do it safely.

10.2.3. Inadequate health and safety management

If your standard of managing health and safety is deemed to be far below what's required and gives rise to a significant risk, inspectors should prosecute. No examples are given in the HSC Enforcement Policy Statement, but it's suggested that a management failure, e.g. deliberately turning a blind eye to unsafe practices on the shop floor in order to maintain production levels would merit prosecution.

10.2.4. Obstruction of inspectors

This would happen when an inspector has been deliberately obstructed from going about his business on an inspection or accident investigation. If an inspector is assaulted whilst going about his duties, the authorities have a policy of seeking police assistance to prosecute the offenders.

10.3. REQUIREMENTS ON ENFORCING AUTHORITIES

The new policy makes a number of requirements on inspectors whilst carrying out their roles and these are given below.

10.3.1. The role of management

They must consider the role of the management chain and that of individual directors and managers in any possible offences. So if evidence suggests that enforcement action should be taken against particular individuals, then that should happen. Where appropriate, enforcing authorities should seek disqualification of directors under the legislation that governs the disqualification of company directors.

10.3.2. Publicising enforcement actions

The names of all organisations and individuals convicted of a health and safety offence during the previous twelve-month period must be published. Similar information regarding all Improvement and Prohibition Notices issued over the same period should be provided.

11. Useful addresses

If you don't know where to look for help with health and safety, you're certainly not alone. We've included a few addresses, with a brief description of how these organisations can help you.

11.1. BRITISH SAFETY COUNCIL

The British Safety Council is a large organisation concerned with health, safety and environmental issues. Its website has some free information, but the real benefits of access to their information service and safety network will be for organisations who subscribe and become members. Members receive other benefits such as a helpline and a magazine.

Tel: 020 8741 1231

Fax: 020 8741 4555

E-mail: www.britishsafetycouncil.org.

11.2. HSE BOOKS

This is the publishing arm of the HSE and is the point of contact for all publications, including multimedia.

The website www.hsebooks.co.uk has a search facility to allow you to find free and priced publications.

Tel: 01787 881165

Fax: 01787 313995

11.3. HSE INFOLINE

This is a very useful, free and confidential service, whereby you can contact the HSE via telephone, fax or email for advice on health or safety problems. It's also an enquiry point for advice on HSE publications.

You won't be asked for any personal details such as your company name and where you're based. The only questions will be general to enable your question to be answered.

Tel: 08701 545500 (open 8.30am-5.00pm, Monday – Friday)

Fax: 02920 859260

E-mail: **hseinformationservices@natbrit.com.**

11.4. INSTITUTION OF OCCUPATIONAL SAFETY & HEALTH

This is one of the leading bodies in occupational health and safety with over 25,000 members. The website contains much free information about different health and safety topics and it has a discussion forum that non-members can access. The advantage of this is that you can see questions and answers to problems that have already been raised by experienced health and safety advisors.

The website www.iosh.co.uk contains a section that enables you to seek a specialist consultant. The service is free and you will need to fill out an online form and submit it to IOSH. They will then forward it to those whom they consider suitably qualified, who will then contact you.

Tel: 0116 257 3100

Fax: 0116 257 3101

11.5. ROYAL SOCIETY FOR THE PREVENTION OF ACCIDENTS

This organisation is commonly known as RoSPA. It can be contacted for advice on a number of safety issues and organisations that you are able to join. One of the current hot topics is driver safety, although RoSPA focuses on health and safety in general.

Its website www.rospa.org.uk contains free information including fact sheets.

Tel: 0121 248 2000 (general information)

Fax: 0121 248 2001

Section II

Fire Authorities

This section provides guidance for businesses in preparing for and dealing with any fire safety inspections that may occur. It's based on the experience of those with knowledge of fire safety inspections in various premises, as well as those who've been on the receiving end of such inspections.

1. Overview

Like the section on health and safety enforcement, references to specific pieces of legislation have been kept to a minimum, as the purpose is to describe what you must do within the context of the powers that inspectors have when they visit. The emphasis is very much on a practical approach that's realistic for a small business and is one that should help you to "manage" the visit more effectively.

This section also provides information on how premises are selected for inspections, through to what an inspector will expect to see on his visit. Information is also given on typical problems that can be encountered by businesses and tips are provided on how to avoid them.

Much emphasis is given to the importance of paperwork, not only so you can demonstrate that you've done what the law requires, but because the right paperwork can help you establish a good defence should the need arise.

2. Who undertakes inspections?

2.1. WHO WILL VISIT YOU?

Fire safety inspections are undertaken by three primary authorities depending on the type or use of a property and the particular legislation that applies to it.

2.1.1. Fire authority

The local fire authority is the safety branch of the fire brigade and it is responsible for inspecting the majority of premises.

2.1.2. HSE

The Health & Safety Executive (HSE) is responsible for fire safety on construction sites and other special premises such as mines and quarries. However, this shouldn't be confused with its main role of "factory inspector" which is predominantly for health and safety enforcement.

2.1.3. Local council enforcement

Depending on the nature of your premises, a number of different council officials could be involved. So, if your premises are houses in multiple occupation such as nurses' homes or halls of residence, you will come under the responsibility of the local council, Metropolitan County Authority Housing Officer or Environmental Health Officer. Whereas, if you run a residential elderly care, nursing or children's home the council's Social Services department will cover you. Though this may depend on the arrangements locally. In most of these situations the local authority departments have "agency" arrangements with the corresponding fire authorities and use the fire authority's advice when inspecting premises.

2.2. DIFFERENT APPROACHES

Such a diverse range of inspectors can lead to a different approach or application of the relevant legislation. Fire Officers are immersed in mainstream fire safety practice and normally have a wealth of broad experience. HSE inspectors are usually highly qualified safety practitioners but, like qualified local authority officers, may have little in the way of practical experience of fire safety matters. In these cases they almost always rely on the available guidance when making recommendations.

3. Selection criteria

3.1. SELECTING PREMISES

There are basically six reasons why an authority will undertake an inspection to assess fire safety provisions in your premises.

3.1.1. To review premises

This may be to review premises where an application has been made for a new or amended fire certificate. This could be as a result of the premises having works in progress or proposed works for which Building Regulations approval has been sought, and where the local building control authority is obliged to consult with the fire authority.

3.1.2. Complaint

Sometimes the authorities will receive a complaint from a third party expressing concern about fire safety on your premises. This could be a visitor or even a member of staff.

3.1.3. Routine inspection

Routine inspections are very much the "luck of the draw" and depend on the nature of your premises and the enforcement priorities of your local inspectors. These types of inspection are often used to review compliance with a current fire certificate for those premises that have them. They are also carried out on exempt premises, to see if they should continue to be exempt and that general fire safety is satisfactory.

3.1.4. Workplace inspections

This is to ensure that a mandatory risk assessment has been undertaken.

3.1.5. Agency arrangements or on-loan

Such visits arise as part of an "agency" arrangement with another enforcement body. This occurs where the other body is using the expertise of the fire inspector to assist in the assessment of the premises for other purposes, e.g. registered care homes.

3.1.6. After an emergency

Inspectors are likely to visit during or following a fire incident.

3.2. FREQUENCY OF INSPECTIONS

The frequency of inspections varies greatly across the UK, as there's no mandatory requirement under which authorities must make inspections at a pre-determined frequency. However, it's likely that more inspections will be made to those premises that have had problems in the past and to those yet to rectify previous problems.

4. Powers of entry

All inspecting authorities have rights of immediate entry onto premises within their jurisdiction, though most have the courtesy to announce their visits in advance. This usually works in their favour by ensuring occupiers have all relevant information and paperwork to hand. If they don't announce their visit they would normally expect to be admitted unless you can show good cause why it should be re-arranged. Sufficiently good reasons for postponement will vary between fire officers, but you would need to demonstrate an adverse effect on business such as severe short staffing or that a major order will be lost.

4.1. COOPERATION IS BEST

If you're really busy and an inspector just turns up, see if you can give them a brief walk around and postpone the full inspection.

It should be noted that most authorities undertake fire inspections in an effort to foster good relations with businesses and whilst it may be considered an inconvenience for them to attend unannounced, you should make every effort to enable the inspection to take place at the time. This will avoid any unpleasantness that a lack of cooperation may suggest and will avoid the inspector getting the impression that you may have something to hide. If you're genuinely hard pressed for time, try negotiating for a brief inspection by offering a quick walk around the premises. This may make it easier for you to arrange a more convenient time for the inspector to return in order to carry out a full inspection.

5. The inspector's wish list

A fire inspection will consist of different tasks depending on the type and size of premises and the nature of the work processes involved e.g. an inspection of offices will be quite different from a visit to a factory that uses highly flammable chemicals. However, any inspection will almost certainly involve a walk-through of the buildings, a discussion with key employees and a review of documents.

5.1. FIRE SAFETY ESSENTIALS

Your inspector will be particularly interested in the means of escape from your building such as the corridors, stairs and final exit points to street level. These may include the fire alarm system, any smoke detection equipment connected to it, signs and notices, emergency lighting, fire extinguishers, sprinkler systems (where they are fitted), door fastenings and training and procedures.

An inspector will expect to see a fire risk assessment, even in buildings with fire certificates.

5.2. COMPLIANCE WITH FIRE CERTIFICATE

Where a fire certificate has been issued it details all the fire safety features present in the building. So an inspection to review compliance with it may be limited to simply walking around the building and checking-off the plan accompanying the certificate with the actual layout and content of the building. Here, it's easy for you to ensure that the premises show compliance with the certificate as the exact information as to what's required is stated on it.

Where no certificate has been issued or where the premises aren't required to have one, assessment of fire safety is less objective. This is why it's essential that all premises in which people work have a current fire risk assessment. Note that a fire risk assessment is also required in those premises that do have a fire certificate.

5.3. FIRE SAFETY CHECKLIST

The following is a brief but not exhaustive checklist of what fire inspectors will want to see and possibly discuss with you on their visit.

A number of fire brigades offer very reasonably priced courses that are much cheaper than more commercial providers e.g. a one-day course on risk assessment costs around £100 per person.

5.3.1. Fire safety risk assessment

These are required by specific fire safety legislation and must be in written form if five or more people are employed. In practice, they are similar to normal risk assessments but are specifically concerned with fire safety provision.

5.3.2. Fire certificate

The fire certificate is one of the first things that an inspector will ask to see on premises where one is a requirement.

5.3.3. Records

The inspector will want to see the following records: routine testing and maintenance of the fire alarm (where fitted), sprinkler system, hose reels or fire fighting risers, emergency lighting systems, fire extinguishers and smoke extraction fans.

5.3.4. Fire safety training

This includes details of the fire safety training given to staff, including names of those attending, the dates of the sessions and training subject details. You may also be asked about the qualifications of the trainer giving the course and their experience.

5.3.5. Fire drills

You should have details of any routine fire drills that you've undertaken to hand. These details should include the dates of the drills and notes of any problems that may have occurred during the evacuation e.g. a fire alarm failing to go off, or staff failing to evacuate. However, if some problems have been noted, an inspector is likely to ask about them and the measures that you have put in place to prevent a recurrence.

5.3.6. Special measures

Some premises have goods stored on site that present a greater fire risk and as a result extra precautions need to be taken. For example, you'll need to show evidence of any special measures required with regard to the storage of high-risk substances e.g. locked, ventilated cabinets for flammable liquids. In these circumstances, it's also useful to be able to demonstrate that you keep the minimum volume possible on site, as stockpiling such liquids poses an unnecessary risk.

5.3.7. Extra maintenance requirements

This depends very much on your own particular workplace. But if you do have extra maintenance needs, you're likely to be asked for details and records of any that occur. This is to ensure that any high-risk equipment or processes are maintained safely e.g. the use of portable welding equipment.

5.3.8. Walkabout

Inspectors may question your staff on where the fire exits are, so make sure everyone knows.

Inspectors will want to walk around the building and expect to be given access to all areas. They may ask questions of employees on the way, such as when the last fire evacuation took place, where the nearest emergency exits are and what they would do if they discovered a fire.

They'll expect to be given a brief commentary of what goes on in each area and may ask you to state what you perceive to be high-risk. They will expect to see that all systems relating to fire safety are in place and working correctly and an explanation where this is not the case.

6. Problems encountered (and how to resolve them)

Whilst there is no such thing as a "typical" non-compliance when it comes to fire safety inspections, a variety of problems can occur. These range from those which are most serious and immediately life-threatening, to those which are inevitable and which usually occur with age, wear and tear and the normal life-cycle of a building.

6.1. CHANGES TO THE BUILDING LAYOUT

6.1.1. Day-to-day changes

In premises where a fire certificate has been issued the most commonly occurring fault is when its "conditions" have been breached. This is usually associated with day-to-day changes in the layout of the building.

6.1.2. Starting from scratch

Make life easier for yourself and talk to the fire brigade whenever you wish to make any type of change.

This happens when a building has been adapted or added to over the years, but the fire certificate hasn't been upgraded to reflect the changes. This can happen and when it does it will be the worst time to try and get the certificate altered. This is because the inspecting authority will often start from scratch when assessing the fire safety measures that are required. It is situations like this that emphasise the importance of consulting with the fire authority over any changes to building layout, type of occupancy or even introducing a new work process that may have fire safety implications.

Any one of these changes may mean that a new fire certificate will require additional fire protection measures such as a new fire alarm, automatic detection equipment or new escape routes. However, this scenario is fairly rare and usually occurs when employers have been negligent in not informing the authorities of such changes.

6.2. MINOR CHANGES

In the majority of cases it's minor matters where fire certificates have been contravened, and these include missing fire exit signs, e.g. where walls have been decorated, fire extinguishers used as coat hangers (thus hiding their presence), fire doors propped open to allow ventilation of stuffy rooms, and partial obstruction of escape routes used ordinarily for storage. However, these are common faults and unless an inspector has warned you about them repeatedly on previous visits, you can expect a verbal comment to sort the faults out. Most are easy to resolve at the time or within a reasonable period at low cost and with minimal disruption to the business.

6.3. RISK ASSESSMENT

It's worth noting that as well as safeguarding the business against enforcement action by fire and other authorities the provision of an adequate risk assessment will help to protect you. This will be against property and other loss or the exposure to private prosecution by employees or visitors. In an era of soaring Employer's Liability Insurance premiums, no fire risk assessment may mean no insurance cover.

No fire risk assessment may mean failure to obtain or renew Employer's Liability Insurance.

Figures from the Association of British Insurers (ABI) show that the cost of commercial property fire claims rose by 30% from 2000 levels to £679 million in 2001. As a result the ABI is warning that greater emphasis on fire prevention is needed in order to counter the impact of current increases in insurance premiums. So risk assessments may also help you keep increases in your premiums down at renewal time.

6.3.1. Go back to basics

There are still many employers that have failed to undertake a fire risk assessment and where this is the case, the fire authority has a number of options (see later). But the risk assessment itself need not be over-complex or exhaustive. It simply needs to demonstrate that you are aware of the risks to which people are exposed, and have taken reasonable steps to reduce them to a minimum. The inspecting authority will merely wish to check that you have undertaken this process.

6.3.2. A useful defence

Where there may be conflicts with older legislation, risk assessments can be used to justify that your building is safe.

Example 1

A room may be shown on the fire certificate as needing a fire resistant, self-closing door to prevent fire and smoke passing from it to the adjacent escape route. However, during normal use the room is staffed, but extremely stuffy so the fire door is propped open. Whilst this is a direct contravention of the conditions of the fire certificate (which states that fire doors must be kept closed at all times) the business may be able to demonstrate, through a risk assessment, that the resulting propped-door is not unsafe.

You may be able to argue that there are procedures that whenever the room is unoccupied, for whatever length of time, the door is closed. Whilst occupied by staff there is detection far better than any smoke detection on the market, and in the event of a fire the occupants would leave, close the door and raise the alarm. Thus the propping of the door in no way adversely affects fire safety provided the occupants continue to manage and implement the procedure. This is called a "safety case".

Example 2

A fire exit sign provides advice to occupants on the exits available and is normally used to highlight exits that are not normally used (e.g. other than the main entrance). During refurbishment works the sign has been removed and not replaced. This again is a direct contravention since a sign is shown on the fire certificate. However, through risk assessment the employer could demonstrate that the property is occupied only by those predominantly familiar with their surroundings and that, because of general health and safety requirements, any visitors are confined to the main reception or are otherwise always supervised by staff.

The employer could show in the risk assessment and attendant records that bi-annual fire-drills are undertaken involving all staff and all fire exits so all staff therefore know of the routes to each. In the event of an emergency all staff know where the exits are and so the signs are unnecessary. Provided the records of the fire drills are genuine and that staff actually do know where the exits are, the risk assessment procedure would show there is no need for a sign.

6.4. HAVE PROPER MAINTENANCE

Fire alarms, emergency lighting and other safety systems breakdown from time to time and there's often little that can be done to prevent this. Normal cyclical maintenance will reduce the occurrence of breakdowns through wear and tear and if this has been carried out effectively it will be a valid defence.

If a poor excuse is given at the time of an inspection, e.g. the fire alarm system broke down a few days ago and that you've been meaning to call

Have a logbook to show the date of any breakdown and details of call-outs and follow-ups where applicable.

an engineer but haven't got around to it yet, a serious breach of fire safety might be dealt with in a severe manner by the authority. A logbook, however, showing the date of the breakdown and a timely request to the maintenance company, with a follow-up asking why they haven't attended yet, is likely to be met with understanding and support. Simple.

7. Enforcement action and appeal

7.1. THE NOTICES

7.1.1. "Steps to be taken" Notice

If contraventions are found in premises where a fire certificate has been issued, the authority may issue a **"Steps to be taken" Notice**, requiring defects to be rectified or remedial measures to be put in place within a specific time period. This depends on the defect noted, though there is room for discussion on any time scales.

7.1.2. Improvement Notice

In premises where a fire certificate is not required but where the legislation applies anyway (this is most workplaces), an **Improvement Notice** may be issued, which again specifies the required remedial works within specific time periods. These periods will vary depending on what needs to be done but could be several months.

7.1.3. Enforcement Notice

In all premises where fire risk assessment is mandatory (all workplaces), an **Enforcement Notice** can be issued which stipulates what measures the authority considers necessary to reduce risks to a reasonable level. Time periods for complying are variable and like Improvement Notices are subject to some negotiation with the fire authority.

7.1.4. Prohibition Notice

In any building where the risk of death or injury through fire is very serious, the fire authority can issue a **Prohibition Notice**, which restricts use of part or all of the building, until the problem is rectified. The time period will depend on each individual case and the potential risk.

7.2. DISCUSSION TIME

Except in the most severe cases, authorities will issue Notices requiring work to be done within a period of time, and they will normally discuss their intentions prior to issuing the Notice although this is not a statutory requirement. They will almost always seek to obtain your consent to the works, and even discuss optional timescales for its completion, prior to formally issuing the Notice. The reason for this is simple, to avoid unnecessary court time.

7.3. COMMERCIAL NEEDS v SAFETY

In the majority of cases fire authorities and inspecting officers recognise the normal pressures under which a business has to operate, and will show discretion when imposing timescales. However, they will also be mindful of the responsibility of an employer to safeguard people in and around the building and should not be expected to compromise standards in favour of financial ability to undertake works.

Where remedial works are necessary, you do have room to negotiate with the authorities.

This is where you will need to carefully consider and justify exactly how much time and cost is involved in the remedial works. You will need to assess what the benefit of reducing overall risk might be by undertaking a larger amount of lower cost work in a shorter time period and then possibly negotiating a longer timescale for more costly or disruptive elements.

7.4. LOW-COST MEASURES

Remedial measures don't have to be costly, so think about what you can do through procedures and management (often at zero or low cost) to improve the situation, e.g. moving a large number of desk-based employees to the ground floor and storage to an upper floor and increasing inspection frequency of equipment.

7.5. TIME PERIODS

Generally, the most common time-frame for remedial works is anywhere between three and six months. Where extensive work may be required, periods of up to two or three years have been known. However, it's to be expected that simple measures, such as additional fire exit signs or fire extinguishers, should be expected to be in place within days.

7.6. POTENTIAL CONFLICT

If you have failed to undertake a risk assessment, or if the authority considers that the risk assessment and its control actions are inadequate, they can issue a Notice requiring that certain measures be undertaken.

7.7. APPEAL PROCESS

The appeal procedure remains the same, in that you have 21 days to apply to a Magistrates' Court (Sheriff's Court in Scotland) for the appeal to be heard. In all cases except where a Prohibition Notice has been issued, the lodging of an appeal suspends any time limits put in place by any Notice served by the authority. Any period or any revised period for compliance with the Notice that's designated by the court will only take effect after determination of the case. However, Prohibition Notices remain in force until the court's decision.

7.8. ALTERING OR WITHDRAWING NOTICES

It should also be noted that the fire authority can withdraw or alter a Notice after it has been issued at any time, at its discretion, or extend the time stipulated on it. If a Notice has been issued it's therefore wise to remain in contact with the authorities, informing them of progress as it develops, since early warning of delays is more likely to lead to the authority supporting and granting an extension.

8. Getting the most out of the visit

8.1. RAISE THOSE CONCERNS

Bringing matters to the attention of the inspector that you know to be a problem is, from experience, a wise move. If you know about a deficiency you are obliged by law to rectify it anyway, and to hope that the fire officer will go around without noticing is rather tempting fate! Up-front honesty is more likely to encourage assistance.

8.2. KEEP TALKING

Some fire authority inspectors are the silent types who say nothing until the end, leaving it to the employer's imagination and interpretation of facial expressions as to how the visit is going. Current trends in interpersonal training encourage authorities to give "positive strokes" by making encouraging comments where things are going well. A tip would be to ask how things are going from time to time if this information is not forthcoming. Another idea is to ask frequent questions about everyday issues, which you find troublesome, such as how to get staff on to a quality fire safety training course, or the problems with people parking near fire exits. Experienced fire officers will have a wealth of anecdotal advice to hand (and it's free!).

Inspectors prefer the up-front and honest approach where employers know that defects are present.

8.3. FEEDBACK TIME

The end of the visit should be a "feedback" session where the officer verbally recounts what he's seen and includes any immediate advice on remedial measures. As stated he should also provide positive comments on those elements that are satisfactory. You would be wise to compliment the officer on the assistance given, provided, of course, that his advice was helpful. Some authorities give a hand-written copy of their findings at the time, but with the increasing demands of quality control and verification, it is more likely that you will have to wait for the written outcome.

If the inspector is the silent type, try and draw him out as to how the visit is going.

9. The complaints procedure

If things didn't go well you can contact the fire authority and make comments concerning your dissatisfaction with the visit. However, whilst enforcement authorities normally take such complaints seriously, you should use it only where absolutely necessary as it isn't a positive step forward in building a satisfactory relationship with the inspector concerned.

9.1. WHO DID THE INSPECTION?

It may be that the inspection was undertaken for other purposes e.g. under Registered Care Homes legislation or of a construction site by the Health & Safety Executive. In these cases, although it's desirable that a fire authority officer accompanies any inspector under an agency arrangement, this may not have been the case. The results of these visits are often vague and frustrating for businesses especially where it is apparent that the officers concerned are not well versed in fire safety and simply quote from readily available guidance documents. In these cases you should contact the fire authority directly and seek to arrange a further joint visit whereby the advice of the "expert" is obtained at first hand. In reality, the inspecting authority often send their comments to the fire authority, who would be well advised to visit the premises before endorsing them, and this is often the case.

9.2. COMPLAIN OR APPEAL?

Beyond complaining to the line manager of the officer concerned there's little else formally available to you, other than the appeals procedure against any formal enforcement action taken. The advice from an independent consultant may be appropriate but you should be aware that the advice might be the same as that required by the authority.

10. Future developments

10.1. GOODBYE FIRE CERTIFICATES

At the beginning of August 2002, it was announced that the government plans to conduct a big shake up of fire legislation with the aim of simplifying it. The biggest factor in this shake up would be the end of many businesses needing to have Fire Certificates. Instead, employers would be made totally responsible for carrying out fire risk assessments on their premises and they would need to implement any risk reduction strategies that the assessment showed was needed. As a rough time frame, it's intended that these new changes will come into force in 2004.

10.2. WHY ARE THEY BEING SCRAPPED?

Surveys suggest that fewer than 60% of employers are aware of their duties under the current legislation. So the main reason for the proposed changes is to remove the problems and to increase compliance rates by having a simplified system. Earlier legislation will either be repealed or amended to remove references to fire safety as such references are contained in many pieces of different legislation, which adds to the confusion. This should help small businesses in particular, especially as the Federation of Small Businesses was involved in developing these proposals.

10.3. NO MORE FEES

The end of Fire Certificates would also mean that you no longer have to pay for a certificate, thus saving businesses around £1.7 million a year. However, there would still be visits from the fire authorities in order to inspect premises and to ensure that fire provision is adequate.

10.4. EXTRA POWERS

It's intended that fire authorities will get extra powers to take action against suppliers of fire alarm equipment that produce regular false alerts. Also fire safety regulation would extend into the voluntary sector and will now affect some self-employed workers, though it's not yet known how. Another aim is to remove much routine work from fire brigades to allow them to focus their resources on inspection and enforcement of higher risk premises. Fire brigades are also likely to be given statutory duties to promote community fire safety.

11. Useful addresses

If you don't know where to look for help on fire safety issues, you're not alone. We've included a few addresses, with a brief description of how these organisations can help you.

11.1. Fire Industry Confederation

This is a Confederation of the four leading UK Fire Protection Trade Associations who collectively represent over 90% of the UK Fire Protection Market. If you are looking for fire extinguishers, sprinkler systems or any other type of fire protection, details of members can be found on this site.

Tel: 020 8549 8839

Fax: 020 8547 1564

www.the-fic.org.uk.

11.2. Fire Protection Association

This is the UK's national fire safety organisation. Its purpose is to draw attention to the dangers of fire and to advise on fire prevention strategies to minimise losses arising from fires. It's supported by the Association of British Insurers and Lloyd's of London.

The FPA has a consultancy and training division as well as producing a series of publications on fire safety.

Tel: 020 7902 5300

Fax: 020 7902 5301

www.thefpa.co.uk.

11.3. Institute of Fire Prevention Officers

This has a broader base than the Institution of Fire Engineers and has membership from enforcement officers, manufacturers, consultants and advisors, amongst others. It has bulletin boards that are accessible to non-members on every fire safety related subject.

It is a source of advice for anyone who needs help and can direct you to a suitable consultant or company.

IFPO

PO Box 677

Croydon

Surrey CR2 0ZH

No telephone or fax number given

www.fire.org.uk/IFPO/home.htm.

11.4. Institution of Fire Engineers

This is a professional body for those engaged in the field of fire engineering. Like any such body, it seeks to set standards for those wishing to work within fire engineering and does this by having a defined membership structure based on experience and qualifications.

The website has a register of consultants and it's free to search, though it will be down to you to satisfy yourself as to the competence of any consultant selected.

Tel: 0116 255 3654

Fax: 0116 247 1231

www.ife.org.uk.

11.5 Loss Prevention Council

The Loss Prevention Council (LPC) is owned by the Association of British Insurers and Lloyd's of London. It provides a variety of technical services in the field of loss prevention (including research), standard setting, product testing, certification and training.

Tel: 020 8207 2345

Fax: 020 8207 6305

www.lpc.co.uk

11.6. Your local fire brigade

Look in a telephone directory and ring them up. They offer lots of advice and many offer good quality training courses.

Fire brigade training courses are much cheaper than other commercially run courses and the trainers are usually ex-fire fighters.

Section III

Checklists

This chapter includes a series of checklists to help you comply with health and safety requirements. They are divided into different areas, so they can be mixed and matched, depending on the needs of your own particular workplace. A checklist is included to help you comply with fire safety requirements. In premises that have a Fire Certificate, it will be the certificate that details what must be done and how, but in other cases, a simple checklist can help on a day-to-day basis, though it's no replacement for a risk assessment.

HEALTH AND SAFETY CHECKLIST: OFFICES

Question	**Y**	**N**
• Is the area tidy with good housekeeping?	❑	❑
• Are rubbish bins adequate and emptied regularly?	❑	❑
• Are items stored properly in cupboards and on shelves?	❑	❑
• Is the flooring level and free of trip hazards?	❑	❑
• Are the floors and walkways clear of obstructions?	❑	❑
• Are the floors clean?	❑	❑
• Is there enough room for staff to work and move around in?	❑	❑
• Is the first aid kit/room stocked with adequate supplies?	❑	❑
• Are the contents of the kits within their expiry dates?	❑	❑
• Is the lighting suitable for the work done in the area?	❑	❑
• Does the emergency lighting work?	❑	❑
• Is glare a problem?	❑	❑
• Is there any light reflected from shiny surfaces?	❑	❑
• Have all electrical appliances, leads and plugs been tested?	❑	❑
• Are extension leads in good working order?	❑	❑
• Are there any broken switches or power outlets?	❑	❑
• Is there any damage to cupboards, shelves or other fittings?	❑	❑
• Is the furniture in good condition e.g. no broken chairs?	❑	❑

HEALTH AND SAFETY CHECKLIST: GENERAL WORK AREA/WORKSHOP

Question	Y	N
• Is the area tidy with good housekeeping?	❑	❑
• Are workbenches or work surfaces clean and tidy?	❑	❑
• Are the floors clean?	❑	❑
• Is the flooring level and free of trip hazards?	❑	❑
• Are the floors and walkways clear of obstructions?	❑	❑
• Are any openings covered or secured to prevent access?	❑	❑
• Is there enough room for staff to work and move around in?	❑	❑
• Are aisles adequate, passable and clearly marked?	❑	❑
• Are emergency exits clearly marked and free of obstruction?	❑	❑
• Are machine guards provided and used?	❑	❑
• Is there adequate spacing between machinery?	❑	❑
• Are the machines clean?	❑	❑
• Does the noise level interfere with normal conversation?	❑	❑
• Has the noise level been measured?	❑	❑
• Do workers need hearing protection?	❑	❑
• If so, are the earplugs/ear defenders properly maintained?	❑	❑
• Have workers been trained how to use the hearing protection?	❑	❑
• What engineering controls are in place to reduce noise levels?	❑	❑
• What is the general condition of tools?	❑	❑
• Are guards in place and secure?	❑	❑
• Do workers know how to use the tools?	❑	❑
• Is there a system for repair and replacement of tools?	❑	❑
• Are there drip pans to prevent spillages?	❑	❑
• Is there adequate provision to store waste materials?	❑	❑
• Are there enough suitably stocked first aid boxes?	❑	❑
• Is the safety policy displayed prominently?	❑	❑
• Are the evacuation procedures clear?	❑	❑
• Are there adequate waste bins?	❑	❑
• Are the rubbish bins emptied regularly?	❑	❑

HEALTH AND SAFETY CHECKLIST: CATERING

Question	Y	N
• Are the food preparation areas well-constructed?	❑	❑
• Are they easy to clean?	❑	❑
• Are the storage areas well constructed?	❑	❑
• Are they easy to clean?	❑	❑
• Are all the work surfaces in a good state of repair?	❑	❑
• Are all the work surfaces easy to clean?	❑	❑
• Is all the equipment in a good state of repair?	❑	❑
• Are all the items of equipment easy to clean?	❑	❑
• Are there enough washbasins within reach of food preparation areas?	❑	❑
• Do these have hot and cold running water?	❑	❑
• Are there enough sinks for washing both food and equipment?	❑	❑
• Are any of the food ingredients/dishes likely to become contaminated by chemicals, pests or dirt?	❑	❑
• Are raw and cooked foods stored separately?	❑	❑
• Do any food ingredients/dishes need to be kept refrigerated?	❑	❑
• Could the food be at risk of contamination?	❑	❑
• Does the food reach an internal temperature of $70^{O}C$ for at least two minutes?	❑	❑
• Is the cooked food at risk of cross-contamination after cooking?	❑	❑
• Is the food left to cool at room temperature after cooking?	❑	❑
• Is the food held hot/chilled after cooking?	❑	❑
• Is the food re-heated before it is served?	❑	❑

HEALTH AND SAFETY CHECKLIST: CHEMICALS

Question	Y	N
• Are chemicals in correct and secure storage?	❑	❑
• Are chemicals correctly labelled and packaged?	❑	❑
• Are chemicals kept in shatterproof containers?	❑	❑
• Are chemicals handled and used in accordance with the labels or Material Safety Data Sheets (MSDS)?	❑	❑
• Are staff trained in the correct use and handling of chemicals?	❑	❑
• Are metal parts, paintbrushes etc. cleaned with solvents such as turpentine in open containers?	❑	❑
• Are solvents and flammable liquids stored in a safe manner?	❑	❑
• Is there any chance of isolating work involving chemicals from the general work area?	❑	❑
• Is the general ventilation system adequate?	❑	❑
• Is there a separate extraction system?	❑	❑
• Is the work area kept free of spilled chemicals?	❑	❑
• Are staff aware of proper waste disposal?	❑	❑
• Are staff trained in hazard awareness?	❑	❑
• Are the correct signs in use and clearly visible?	❑	❑
• Is adequate personal protective equipment (PPE) provided as required by the MSDS?	❑	❑
• Are staff trained in the use of PPE?	❑	❑
• Is the PPE adequately maintained?	❑	❑
• Are first aid facilities available?	❑	❑
• Are eye wash facilities available?	❑	❑

HEALTH AND SAFETY CHECKLIST: PERSONAL PROTECTIVE EQUIPMENT

Question	Y	N
• Are safety helmets worn when required?	❑	❑
• Are there signs advising that safety helmets must be worn?	❑	❑
• Is safety footwear worn when required?	❑	❑
• Are there signs advising that safety footwear must be worn?	❑	❑
• Is protective clothing worn when required?	❑	❑
• Is eye protection (e.g. goggles) worn when required?	❑	❑
• Are there signs advising that eye protection must be worn?	❑	❑
• Is hearing protection worn when required?	❑	❑
• Are there signs advising that hearing protection must be worn?	❑	❑
• Are dust masks being worn when required?	❑	❑
• Are there signs advising that dust masks must be worn?	❑	❑
• Are fall arrestor devices being worn when required?	❑	❑
• Are there signs advising that fall arrestor devices must be worn?	❑	❑
• Are gloves being worn when required?	❑	❑
• Are there signs advising that gloves must be worn?	❑	❑
• Is the PPE serviceable?	❑	❑
• Is the PPE being worn correctly?	❑	❑
• Is the PPE readily available?	❑	❑

HEALTH AND SAFETY CHECKLIST: FIRE

Question	Y	N
• Is there a member of management with overall fire safety responsibility?	❑	❑
• Does everyone know what to do in the event of a fire?	❑	❑
• Are staff trained in the use of fire extinguishers?	❑	❑
• Are the evacuation procedures displayed prominently?	❑	❑
• Are the instructions contained on them clear to staff?	❑	❑
• Are emergency exits clearly marked?	❑	❑
• Are staff encouraged to keep their workplaces tidy?	❑	❑
• Is the workplace kept clear of rubbish?	❑	❑
• Are fire drills held at least annually?	❑	❑
• Is the portable fire fighting equipment in place and regularly inspected?	❑	❑
• Is there clear access to any portable fire fighting equipment?	❑	❑
• Is the fixed fire fighting equipment currently inspected?	❑	❑
• Are the fire alarms currently inspected?	❑	❑